Intraocular Lens Power Calculations

Intraocular Lens Power Calculations

H. John Shammas, MD
Clinical Professor of Ophthalmology
University of Southern California, The Keck School of Medicine
Los Angeles, California

Medical Director
Shammas Eye Medical Center
and M/S Surgery Center
Lynwood, California

An innovative information, education, and management company
6900 Grove Road • Thorofare, NJ 08086

Shammas, H. John.
   Intraocular lens power calculations / H. John Shammas.
      p. ; cm.
   Includes bibliographical references and index.
   ISBN 1-55642-652-6 (alk. paper)
   1. Intraocular lenses. 2. Eye--Accommodation and refraction. 3. Refractive index--Measurement.
   [DNLM: 1. Lenses, Intraocular. 2. Refraction, Ocular. WW 358 S528i
2003] I. Title.
RE988 .S469 2003
617.7'524--dc22
                                    2003015874

Printed in the United States of America.

Published by: SLACK Incorporated
                       6900 Grove Road
                       Thorofare, NJ 08086 USA
                       Telephone: 856-848-1000
                       Fax: 856-853-5991
                       www.slackbooks.com

Dedication

To my wife and best friend,
Najwa Mirhij-Shammas, MD
For her love, understanding, and encouragement.

And to my three lovely daughters,
Rania, Maya, and Lina
Wishing them the best in life
And in their medical careers.

Contents

About the Author

H. John Shammas, MD is a clinical professor of Ophthalmology at the University of Southern California, The Keck School of Medicine in Los Angeles, California, and the medical director of the Shammas Eye Medical Center and M/S Surgery Center in Lynwood, California.

As an ophthalmologist, Dr. Shammas has two major areas of interest; the first one is ophthalmic ultrasonography. He wrote his first textbook, *Atlas of Ophthalmic Ultrasonography and Biometry* in 1984, the second one, *Three-Dimensional Ultrasound Tomography of the Eye* in 1998, and edited his third one, *Ultrasound in Ophthalmology,* in 2000. His second area of interest is cataract surgery and how to improve its outcome. In 1991, he co-authored *Color Atlas of Ophthalmic Surgery: Cataracts* with Richard Kratz, MD. The combination of these two areas of interest can only lead to the subject of biometry and IOL power calculations.

Dr. Shammas has published multiple articles in the American and international literature on axial length measurement and intraocular lens power calculations. As an ophthalmic surgeon, patient satisfaction is very important to him, and avoiding errors in IOL power calculations has been a primary focus of his research studies. He self-published a book on the subject in 1996. Lately, he has focused his research on those cases that had previous refractive surgery, and Chapter 18 is dedicated to this subject.

This book represents the latest developments on the subject and the author's personal experience during the past 25 years.

Contributing Authors

Olivier Bergès, MD
Department of Medical Imaging
Fondation Rothschild
Paris, France

Anastasia Di Giovanni, MD
Resident
University Eye Clinic
Chieti, Italy

Shane Dunne, PhD
Director, OTI Research Laboratory
Ophthalmic Technologies, Inc.
Toronto, Canada

Pier Enrico Gallenga, MD
Professor of Ophthalmology and Head, University Eye Clinic
Department of Surgical Sciences
G. d'Annunzio University
Chieti, Italy

Wolfgang Haigis, PhD
Assistant Professor and Head, Biometry Department
University Eye Hospital of Wuerzburg
Wuerzburg, Germany

Kenneth J. Hoffer, MD
Clinical Professor of Ophthalmology
Jules Stein Eye Institute
University of California, Los Angeles
Los Angeles, Calif

Gus Kohn, CRA, COT, ROUB
Clinical Applications Specialist
Innovative Imaging Inc.
Sacramento, Calif

Adriano Mancini, MD
Ophthalmologist/Teacher
University Eye Clinic
Chieti, Italy

Thomas Olsen, MD
University Eye Clinic
Aarhus Kommunehospital
Aarhus, Denmark

Francois Perrenoud, MD
Department of Ophthalmology
CHIC   ·
Créteil, France

Michel Puech, MD
Department of Ophthalmology
CHNO des XV-XX
Paris, France

Kamal Siahmed, MD
Department of Ophthalmology
Centre Hospitalier Universitaire
Rouen, France

Preface

Intraocular lens (IOL) power calculations have become an integral part of the pre-operative cataract evaluation. During the past two decades, formulas used for such calculations have evolved to a high level of accuracy. The formulas are discussed in the first seven chapters with their advantages and their limitations. Besides the commonly used Holladay 1 & 2, SRK/T, and Hoffer Q formulas, two additional advanced formulas are presented by their authors. These are the Olsen and the Haigis formulas.

Axial length (AL) measurement is one of the major factors affecting IOL power calculations. A-scan biometry is described in detail in Chapters 8 through 13. Two new techniques are gaining in popularity and are herein described in detail by their authors. They include Optical Coherence Biometry by Wolfgang Haigis and B-Mode Guided Biometry by Olivier Bergès et al. The use of phakic IOLs to correct refractive errors has added a new difficulty to AL measurement. In Chapter 16, Kenneth Hoffer describes in detail how to obtain an accurate AL measurement in these cases. Chapter 17 discusses the corneal power measurement and Chapter 18 discusses the intricate ways to determine the IOL power following corneal refractive surgery.

A perfect cataract surgery has to go beyond the surgical technique; it has to meet the patient's visual needs and expectations. Chapter 19 discusses IOL power selection in emmetropes, myopes, and hyperopes. It also discusses IOL power selection in the presence of a capsular rupture with or without vitreous loss and in children.

An error in IOL power calculations is usually suspected after the surgery when the patient presents with an unexpected induced myopia, hyperopia, or aniseikonia. In certain cases, the error is quite large, necessitating a lens exchange as soon as possible. In Chapter 20, I present cases where errors in IOL power calculations have occurred, followed by a discussion on the different alternatives for management.

IOL formulas will always be limited by the accuracy of the data used. This book is aimed at cataract surgeons and their technicians, to increase their understanding of biometry and IOL power calculations and avoid major errors in the future.

My gratitude goes to:

Thomas Olsen, MD for contributing Chapter 4; Wolfgang Haigis, PhD for contributing Chapters 5 and 14; Shane Dunne, PhD for contributing Chapter 8; Gus Kohn, CRA, COT, ROUB for contributing Chapter 11; Pier Enrico Gallenga, MD, Adriano Mancini, MD, & Anastasia Di Giovanni, MD for contributing Chapter 12; Rhonda G. Waldron, MMSc, COMT, CRA, ROUB, RDMS for contributing the Case Presentations in Chapter 12; Olivier Bergès, MD, Kamal Siahmed, MD, Michel Puech, MD, & Francois Perrenoud, MD for contributing Chapter 15; and Kenneth J. Hoffer, MD for contributing Chapter 16.

Through the years, these authors have greatly contributed to the knowledge and advancement of IOL power calculations. Their contribution is greatly appreciated.

H. John Shammas, MD

Basic Optics for Intraocular Lens Power Calculations

All available formulas aim to calculate the exact power of an intraocular lens (IOL) that will be placed inside the eye after cataract extraction and that will produce postoperative emmetropia. After surgery, the two major refracting elements inside the eye would be the cornea and the IOL. Both act as plus lenses and these curved surfaces will refract the incoming rays of light, focusing them on the retina, thus reaching emmetropia.

Refractive Power of the Curved Surface

Refraction is the bending of light as it travels from one medium to another. The interface between the two media is the refracting surface. The refractive index (n) of a transparent material is the ratio of the speed of light in a vacuum to the speed of light in that material. Because light always travels faster in a vacuum, no material has an index of refraction less than 1.00. The commonly used refractive indices in ophthalmic optics are as follows:

Air	1.000
Corneal anterior surface	1.376
Aqueous	1.336
Crystalline lens	1.420
Vitreous	1.336

When the light travels from Medium 1 (index of refraction of n_1) to Medium 2 (index of refraction of n_2), the power (D) of the spherical refractive surface separating them can be calculated using the formula:[1]

$$D = (n_2 - n_1) / r$$

Where r is the radius of curvature in meters.

Figure 1-1. Primary focal point (PFP) of a (+) lens.

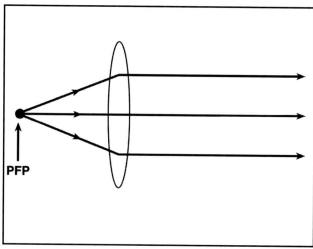

Optics of the Thin Plus Lens

A thin plus lens is characterized by its power, expressed in diopters (D). It indicates the amount of vergence produced by the lens. Each lens has a primary focal point and a secondary focal point. The primary focal point is the point along the optical axis at which an object must be placed for parallel rays to emerge from the lens (Figure 1-1). If parallel rays hit a plus lens, the exiting rays will converge to a specific point; this point is called the secondary focal point (Figure 1-2). The distance between each of the primary and secondary focal points and the center of the lens is the focal distance.[1] The relationship between the lens power (D) and the focal distance (f) in meters is:

$$D = 1/f$$

In Figure 1-3, the object (O) is located in Medium 1 (index of refraction of n_1) at a distance of "u" meters from the lens, and the image (I) is formed in Medium 2 (index of refraction of n_2) at a distance of "v" meters from the lens. The basic vergence formula is:

$$U + D = V$$

Where U is the vergence of the light entering the lens, where $U = n_1/u$, V is the vergence of light exiting the lens where $V = n_2/v$, and D is the vergence of the lens.

The Two-Lens System

In the presence of a two-lens system,[2] it becomes a little bit more difficult to calculate the object-image relationship. The vergences have to be calculated in succession, dealing first with the first lens to encounter the incident light. The image position created by the first lens will then be the object position for the second.

> In the presence of a two-lens system, the vergences have to be calculated in succession, dealing first with the first lens to encounter the incident light.

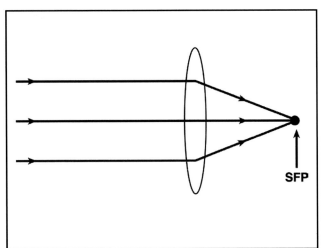

Figure 1-2. Secondary focal point (SFP) of a (+) lens.

SFP

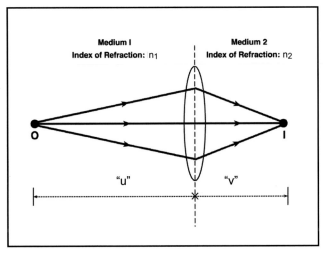

Figure 1-3. Basic vergence of a (+) lens where the object "O" is located in Medium 1 (index of refraction of n_1) at a distance of "u" meters from the lens, and where the image "I" is located in Medium 2 (index of refraction of n_2) at a distance of "v" meters from the lens.

In the operated eye, the incident light is traveling in air (index of refraction of 1.00). These rays first meet the cornea that has a known power (K). The vergence formula for the first lens becomes:

$$U + K = V$$

Since the calculation is for emmetropia, the object is at infinity and the rays are parallel with a zero vergence and U becomes zero; V is the vergence of the image formed by the cornea. Since the image from the first lens system is located at "v" meters from the cornea traveling through the aqueous and vitreous (index of refraction = n), and U is zero, the formula becomes:

$$K = n/v$$
$$\text{And } v = n/K$$

Figure 1-4. The two-lens system as it applies to the operated eye after cataract removal and insertion of an IOL: "c" is the distance between the cornea and the optical center of the IOL, known as the estimated lens position (ELP) or postoperative anterior chamber depth (ACD post); "L" is the distance between the corneal apex and the retina, known as the axial length; "v" is the distance between the cornea and where the image from the first lens (cornea) would have been focused if the IOL had not been inserted.

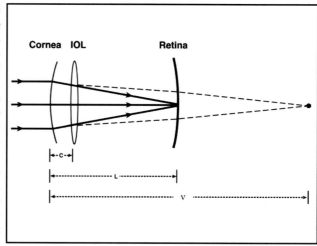

Attention is now shifted to the second lens, which is the IOL that has a power of "P" diopters (D). Figure 1-4 shows that the object for this second lens (which is the image for the first lens) is located at a distance of (v – c), where "v" is the distance between the cornea and where the image from the first lens should be, and "c" is the distance between the cornea and the IOL. The image formed by this second lens is located at the retina if we are calculating for emmetropia. The distance between the IOL and the final image can be calculated by subtracting the distance between the cornea and the IOL, also known as the anterior chamber depth (c), from the distance between the cornea and the retina, also known as the axial length (L).

The vergence formula is written:

$$\frac{n}{v - c} + P = \frac{n}{L - c}$$

It can be rewritten:

$$P = \frac{n}{L - c} - \frac{n}{(n/K) - c}$$

In this formula, P is the power of the IOL needed to produce emmetropia in diopters, L is the axial length of the eye in meters (all formulas adjust the equation to allow the axial length to be entered in millimeters), K is the corneal power in diopters, c is the estimated postoperative anterior chamber depth, also known as the estimated lens position (ELP), and n is the index of refraction of the aqueous and vitreous, which is known to be 1.336.

This basic two-lens formula is at the basis of all theoretical formulas. Each formula introduces adjustments and refinements related to the retinal thickness and the calculation of ELP.

References

1. Smith ME, Kincaid MC, West CE. *Basic Science, Refraction, and Pathology*. St. Louis, MO: Mosby; 2002:87-95.
2. Rubin ML. *Optics for Clinicians*. Gainsville, FL: Triad Scientific Publishers; 1971:27-29.

2

Historic Overview

Before 1975, the power of an intraocular lens (IOL) to be inserted after a cataract extraction was calculated with the use of an equation based on the clinical history:[1]

$$P = 18 + (1.25 \text{ x Ref})$$

Where P is the power of an iris-supported IOL for emmetropia and Ref is the preoperative refractive error in diopters (D) before the development of the cataract.

Errors exceeding 1 D occurred in over 50% of the cases with this clinical history method, and some errors were so large that they were referred to as the "9 D surprise." These large errors were caused by the difficulty of determining the patient's refractive error before the development of the cataract, and the large variations in the crystalline lens power.

A number of formulas for IOL power calculations have since been published; all these formulas are based on an accurate measurement of the corneal power and of the axial length (AL). The original formulas included theoretical and regression formulas. Many of these formulas were then modified in the 1980s to correct the errors in the short and long eyes. Although these formulas are rarely used these days, they are the basis of all modern formulas.

Original Theoretical Formulas

All theoretical formulas for IOL power calculations are based on a two-lens system, the cornea and the IOL, focusing images on the retina.[1]

REQUIRED MEASUREMENTS

The required measurements for calculating the emmetropic lens power (P) and the expected postoperative refractive error (E) are:

(L): The axial length of the eye in millimeters, defined as the distance between the anterior surface of the cornea and the anterior surface of the retina.

(C): The estimated postoperative anterior chamber depth in millimeters, defined as the distance between the anterior surface of the cornea and the anterior surface of the pseudophakos lens.

(K): The corneal power in diopters, or

(R): The corneal radius of curvature in millimeters.

BASIC THEORETICAL FORMULAS FOR EMMETROPIA

Thijssen's,[2] Colenbrander's,[3] Fyodorov's,[4] and van der Heijde's[5] formulas yield approximately the same IOL power for emmetropia. Binkhorst's formula[6] differed by 0.50 D because it was based on a corneal index of refraction of 4/3 instead of 1.3375.

Thijssen's formula:

$$P = \frac{1336}{L - C + Const1} - \frac{1336}{\frac{1336}{K} - C + Const2}$$

Colenbrander's formula:

$$P = \frac{1336}{L - C - 0.05} - \frac{1336}{\frac{1336}{K} - C - 0.05}$$

Fyodorov's formula:

$$P = \frac{1336 - LK}{(L - C)(1 - \frac{CK}{1336})}$$

van der Heijde's formula:

$$P = \frac{1336}{L - C} - \frac{1}{\frac{1}{K} - \frac{C}{1336}}$$

Binkhorst's formula:

$$P = \frac{1336(4R - L)}{(L - C)(4R - C)}$$

RESULTANT REFRACTIVE ERROR

The resultant postoperative refractive error (E) measured at the cornea after the insertion of an IOL with a certain power (I) can be predicted before surgery by Binkhorst's formula[6] or Hoffer's modification of Colenbrander's formula:[7]

Binkhorst's formula:

$$E = \frac{1336 \ (4R - L) \ - \ I \ (L - C) \ (4R - C)}{1336 \ (0.003 \ LR) - I \ (L - C) \ (0.003CR)}$$

Hoffer's formula:

$$E = \cfrac{1.336}{\cfrac{1.336}{\cfrac{1336}{L-C-0.05} - I} + \cfrac{C+0.05}{1000}} - K$$

Original Regression Formulas

Regression formulas are derived empirically from retrospective computer analysis of data on a great many patients who have undergone surgery.

REQUIRED MEASUREMENTS

The required measurements for calculating the emmetropic lens power (P) and the expected postoperative refractive error (E) are:

(L): The axial length of the eye in millimeters, defined as the distance between the anterior surface of the cornea and the anterior surface of the retina.

(K): The corneal power in diopters.

REGRESSION FORMULA FOR EMMETROPIA

A regression formula for emmetropia is based on the following equation:

$$P = A - BL - CK$$

Where P is the IOL power for emmetropia, L is the axial length in millimeters and K is the corneal power in diopters. A, B, C are constants.

The SRK formula[8,9] is the most popular original regression formula where B is 2.5 and C is 0.9. The formula becomes:

$$P = A - 2.5L - 0.9K$$

The constant A varies with the IOL style and manufacturer.

RESULTANT REFRACTIVE ERROR

The formula[8] is:

$$E = 0.67 \ (P - I)$$

Where E is the resultant refractive error, P is the power of the IOL for emmetropia, and I is the power of inserted IOL. The formula is based on data analysis of over 2500 lenses. A change of 1.50 D implant power produces a change of approximately 1.00 D in final postoperative refraction.

Figure 2-1. Calculated IOL power by the SRK and Colenbrander formulas for axial lengths ranging between 20 and 27 mm.

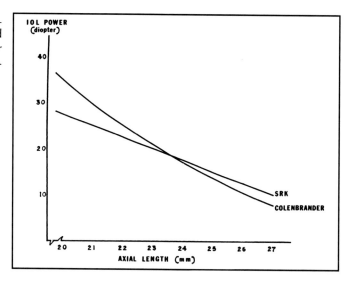

Figure 2-1. Calculated IOL power by the SRK and Colenbrander formulas for axial lengths ranging between 20 and 27 mm.

COMPARING THE SRK REGRESSION FORMULA TO THE ORIGINAL THEORETICAL FORMULAS

Figure 2-1 shows the difference in the calculated power for emmetropia between the original formulas and the SRK regression formula. The values are almost identical when the AL ranges between 23 and 24 mm. The SRK formula measures a stronger IOL power for emmetropia in long eyes and a much weaker IOL power in short eyes.

> The SRK formula measures a stronger IOL power for emmetropia in long eyes and a much weaker IOL power in short eyes.

Modified Formulas for Emmetropia

These formulas are modifications of the original theoretical and regression formulas and were introduced to correct for errors occurring in the long and short eyes.

HOFFER'S FORMULA

Hoffer's formula[10] is based on a modification of Colenbrander's formula. The expected postoperative error at the cornea (R) is added to the corneal power (K). When calculations are made for emmetropia, R becomes nil. Also, the anterior chamber depth (ACD) constant has to be changed in accordance with the AL:

$$\text{corrected } C = (0.292\,L - 2.93) + (C - 3.94)$$

The Hoffer formula is written:

$$P = \frac{1336}{L - C - 0.05} - \frac{1.336}{\dfrac{1.336}{K + R} - \dfrac{C + 0.05}{1000}}$$

Where L is the axial length (mm), C is the anterior chamber depth (mm), K is the corneal power (D), P is the IOL power (D), and R is the refractive error (D) at the cornea.

Shammas' Formula

Shammas' formula[11] is a modified Colenbrander's formula. The first modification involves the corneal index of refraction; it has been changed from 1.3375 to 4/3, as suggested by Binkhorst. The corneal power, such as read on the Bausch and Lomb keratometer (Bausch and Lomb, Rochester NY), is decreased by a factor of 1.0125. The second modification is the incorporation of a fudge factor linked to the AL. This fudge factor has been established by reviewing 500 cases of AL measurements and IOL power calculations in which this factor was needed for higher accuracy. These cases were divided into six categories according to the AL; there is no difference between the calculations and final refractions when the AL was 23 mm. The average difference is +0.25 D for each mm shorter than 23 mm and -0.25 D for each mm longer than 23 mm. Increasing the AL by 0.1 mm for each 1 mm shorter than 23 mm, and decreasing it by 0.1 mm for each 1 mm longer than 23 mm, closes the gap between the calculations and the final refractions in short and long eyes. This fudge factor can be easily introduced into any theoretical formula by changing the axial length (L) to L – 0.1 (L-23).

The Shammas formula becomes:

$$P = \frac{1336}{L - 0.1(L-23) - C - 0.05} - \frac{1}{\dfrac{1.0125}{K} - \dfrac{C + 0.05}{1336}}$$

Where P is the power of the IOL for emmetropia in diopters, L is the axial length in millimeters, K is the average keratometer readings in diopters, and C is the distance between the anterior vertex of the cornea and the IOL in millimeters (estimated ACD).

The Shammas formula is based on AL measurements taken with an immersion technique and does not add any retinal correction. However, in cases where the AL is measured by a contact technique, 0.24 mm should be added to the measurement.

Binkhorst's Adjusted Formula

Binkhorst[12] used his own original formula for emmetropia and varied the postoperative ACD according to the AL.

$$P = \frac{1336 \, (4R - L)}{(L - C) \, (4R - C)}$$

Where P is the power of the IOL for emmetropia in diopters, L is the axial length in millimeters, R is the radius of curvature of the anterior surface of the cornea in millimeters, and C is the distance between the anterior vertex of the cornea and the IOL in millimeters (estimated anterior chamber depth).

The postoperative ACD estimated values (C) are adjusted for iris-fixated, irido-capsular, and posterior chamber lenses. This is based on a positive correlation noted in these cases by Binkhorst between the postoperative anterior chamber depth (C) and the axial length (L).

In the presence of a posterior chamber lens, C is decreased by 0.17 mm for each millimeter the axial length is shorter than 23.45 mm; C is increased by 0.17 mm for each millimeter the axial length is longer than 23.45 mm. The results obtained with Binkhorst's adjusted formula are very similar to results obtained by Shammas' formula.

THE SRK II FORMULA

The SRK II formula[13] is a modification of the original SRK formula:

$$P = A - 2.5L - 0.9K$$

Where P is the implant power in diopters to produce emmetropia, L is the axial length in millimeters, K is the average keratometric readings in diopters, and A is the specific constant for each lens type and/or manufacturer.

In eyes with a normal AL ranging between 22.0 mm and 24.5 mm, there is no change in the SRK formula. In the short eyes measuring less than 22.0 mm, the SRK formula predicts a lower emmetropic IOL power than is actually true. It is corrected by:

- adding 1 D to the calculated P if the AL ranges between 21 and 21.9 mm
- adding 2 D to the calculated P if the AL ranges between 20 and 20.9 mm
- adding 3 D to the calculated P if the AL is under 20 mm

In the very long eyes, measuring over 24.5 mm, the SRK formula predicts a slightly higher emmetropic IOL power than is actually true. It is corrected by subtracting 0.5 D from the calculated P value.

The SRK II formula for ametropic calculations or to calculate the postoperative refraction has been rewritten to read:

$$I = P - (Rt \times Rf)$$

Where I is the implant power in diopters for a target refraction, P is the implant power in diopters for emmetropia, Rt is the desired postoperative refraction, and Rf is the refractive factor.

In the SRK formula, Rf is 1.25 if the power for emmetropia exceeds 14 D and 1.0 if the power for emmetropia is 14 D or less.

Other modifications of the SRK formula have been developed to improve its accuracy. They include the Thompson, Maumenee, and Baker formula[14] and the Donzis, Kastl, and Gordon formula.[15] These formulas are rarely used.

References

1. Shammas HJ. *Atlas of Ophthalmic Ultrasonography and Biometry*. St. Louis, MO: CV Mosby Co.; 1984:273-308.

2. Thijssen JM. The emmetropic and iseikonic implant lens: Computer calculation of the refractive power and its accuracy. *Ophthalmologica*. 1975;171:467-486.
3. Colenbrander MC. Calculations of the power of an iris clip lens for distance vision. *British J Ophthalmol*. 1973;57:735-740.
4. Fyodorov SN, Galin MA, Linksz A. Calculation of the optical power of intraocular lens. *Invest Ophthalmol*. 1975;14:625-628.
5. van der Heijde GL. The optical correction of unilateral aphakia. *Trans Amer Academy Ophthalmol Otolaryngol*. 1976;81:80-88.
6. Binkhorst RD. The optical design of intraocular lens implants. *Ophthalmic Surg*. 1975;6:17-31.
7. Hoffer KJ. Accuracy of ultrasound intraocular lens calculation. *Arch Ophthalmol*. 1981;99:1819-1823.
8. Sanders DR, Kraff MC. Improvement of intraocular lens power calculation using empirical data. *American Intra-Ocular Implant Society Journal*. 1980;6:263-267.
9. Sanders DR, Retzlaff J, Kraff MC. Comparison of empirically derived and theoretical aphakic refraction formulas. *Arch Ophthalmol*. 1983;101:965-967.
10. Hoffer KJ. Intraocular lens calculations: The problem of the short eye. *Ophthalmic Surg*. 1981;12:269-272.
11. Shammas HJ. The fudged formula for intraocular lens power calculations. *American Intra-Ocular Implant Society Journal*. 1982;8:350-352.
12. Binkhorst RD. *Intraocular Lens Power Calculation Manual: A Guide to The Author's TICC-40 Programs*. 3rd Ed. New York: R.D. Binkhorst; 1984.
13. Sanders DR, Retzlaff J, Kraff MC. Comparison of the SRK II formula and the other second generation formulas. *J Cataract Refract Surg*. 1988;14:136-141.
14. Thompson JT, Maumenee AE, Baker CC. A new posterior chamber intraocular lens formula for axial myopes. *Ophthalmology*. 1984;91:484-488.
15. Donzis PB, Kastl PR, Gordon RA. An intraocular lens formula for short, normal and long eyes. *Contact Lens Assoc Ophthalmol J*. 1985;11:95-98.

3

Modern Formulas for Intraocular Lens Power Calculations

The modern theoretical formulas are more complex than the original and modified formulas. The most striking difference is the manner in which the estimated lens position (ELP), also known as the estimated postoperative anterior chamber depth (ACD), is calculated:

- In the original formulas, ELP is a constant value.
- In the modified formulas, ELP varies with the axial length (AL). It decreases in the shorter eye and increases in the longer eye.
- In the modern formulas, ELP varies, not only with the AL, but with the corneal curvature as well. The anterior chamber is deeper in the presence of a steep cornea and shallower in the presence of a flat cornea.

It is important to note that the postoperative ACD does not correlate with the preoperative ACD. Instead, it correlates with the placement of the intraocular lens (IOL), whether it is in the anterior chamber, in the sulcus, or in the capsular bag. The ELP also varies with the implant's configuration and the location of its optical center. The use of a meniscus lens with its anteriorly located optical center calls for a smaller value than a biconvex IOL, where the optical center is more posteriorly located.

> The postoperative ACD does not correlate with the preoperative ACD. Instead, it correlates with the IOL placement within the eye.

The advent of the SRK regression formula brought with it the concept of the A-constant. The A-constant varied not only with the ACD, but also with a variety of factors that had not been taken into consideration in the original theoretical formulas, namely variations in the AL measurement, in the IOL design, and in the surgical techniques. The concept of individualizing the A-constant revolutionized IOL power calculations, leading to the new generations of IOL formulas.

Figure 3-1. In a pseudophakic eye, ELP = aACD + S where: ELP, the estimated lens position, is the distance between the corneal vertex and the IOL's optical center; aACD, the anatomical anterior chamber, is the distance between the corneal vertex and the iris plane of the pseudophakic eye; and S, the surgeon factor, is the distance between the iris plane and the IOL's optical center.

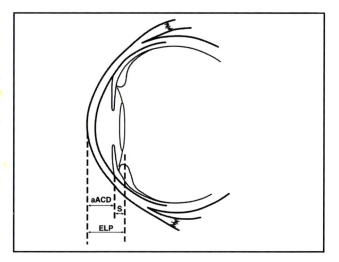

The best known modern formulas are the Holladay formulas,[1,2] the SRK/T formula,[3] and the Hoffer Q formula,[4] and we will confine our discussion to these three formulas. Each formula comprises multiple equations and instead of listing these equations, we will relate the rationale of these formulas. These formulas are readily available on calculators, computer diskettes, and ultrasound units.

Holladay and Holladay 2 Formulas

Holladay's formula[1] is a modification of the theoretical formula based on a three-part system that includes data screening criteria to identify improbable AL and keratometry measurements; a more accurate postoperative ACD estimate that increases the accuracy in short, medium, and long eyes; and a personalized "surgeon factor" that adjusts for any consistent bias in the surgeon's formula.

THE DATA SCREENING CRITERIA

The data screening identifies measurements that are unusual and might require remeasurement, such as:
- An AL of less than 22.0 mm or over 25.0 mm.
- An average corneal power of less than 40 diopters (D) or over 47 D.
- A calculated emmetropic IOL power of more than 3 D from specific lens style.
- A difference between the two eyes of over 1 D in the average corneal power, of over 0.3 mm in the AL, and of over 1 D in the emmetropic IOL power.

A MORE ACCURATE POSTOPERATIVE ANTERIOR CHAMBER DEPTH

The estimated postoperative ACD, also known as the ELP, is the sum of the anatomic ACD (aACD) and the distance from the anterior plane of the iris to the optical plane of the IOL, known as the surgeon factor (S) (Figure 3-1):

$$ELP = aACD + S$$

The aACD is the distance from the corneal vertex to the anterior iris plane after surgery. This distance is more accurately predicted by using a mathematical formula based on the corneal curvature and the axial length:

$$aACD = 0.56 + R\,[R^2 - (AG^2)\,(1/4)]^{-2}$$

Where R is the corneal radius of curvature and AG is the anterior chamber diameter calculated by the equation AG = Axial length x 12.5 x (1/23.45).

A PERSONALIZED SURGEON FACTOR

Although theoretically the S factor is a measurable distance, it is calculated by solving the formula in reverse, using as input variables the postoperative corneal and AL measurements, the IOL power implanted, and the stabilized postoperative refraction.

Just like the A-constant, the S factor represents variations due to lens style, lens manufacturer, surgeon's technique, and measurement devices. The relationships between the S factor, the A-constant, and the ACD value are given by Holladay:

$$S\ factor = (A\text{-}constant \times 0.5663) - 65.60$$

$$S\ factor = (ACD\ value \times 0.9704) - 3.595$$

The S factor represents variations due to lens style, lens manufacturer, surgeon's technique, and measurement devices.

Due to the close relationship between the S factor and ACD value (almost 1:1), a change of 1 unit in the S factor is identical to a change of 1 mm in the ACD value and affects the postoperative refraction in an average eye by around 1.5 D.

The most commonly used S factors are as follows:

Anterior chamber lens	-0.7 to -0.4
Iris supported lens	-0.1 to +0.1
Posterior chamber lens in the sulcus	
Plano-convex	+0.1 to +0.3
Biconvex	+0.4 to +0.7
Posterior chamber lens in the bag	
Plano-convex	+0.9 to +1.1
Biconvex	+1.2 to +1.6

A 1-unit change in the S factor affects the postoperative refraction by around 1.5 D.

HOLLADAY 2 FORMULA

The Holladay 2 formula attempts to predict a more accurate ELP and bases its ELP calculations not only on the axial length and K readings as in the original Holladay formula, but it also takes into account the white-to-white corneal measurement, the phakic ACD, the phakic lens thickness, and the patient's age and sex.

The Holladay 2 formula has never been published (to the author's best knowledge), but it is available as part of the *Holladay IOL Consultant* (Holladay LASIK

Institute, Houston, TX), a computer program that also includes the following features:[2]
- Standard IOL power calculations in the phakic, aphakic, and pseudophakic eyes.
- Axial length measurement corrections in aphakia and pseudophakia.
- Multiple formulas for comparison.
- An IOL database.
- An alternate K value calculation for post-keratorefractive surgery.
- Lens constant personalization.
- Other features include outcome analysis, back-calculations, prediction error report, astigmatism analysis, and potential outcome report.

The SRK/T Formula

The SRK/T formula,[3] contrary to the SRK and the SRK II formulas, is a theoretical formula based on Fyodorov's formula and uses empirical regression methodology for optimization of the postoperative ACD prediction, of the retinal thickness correction factor, and of the corneal refractive index.

OPTIMIZATION OF THE POSTOPERATIVE ACD PREDICTION

The estimated postoperative ACD is calculated:

$$ACD \text{ (estimate)} = \text{corneal height (H)} + \text{offset}$$
$$\text{Offset} = ACD \text{ (constant)} - 3.336$$

It is based on the hypothesis that the IOL lies at a constant distance from the calculated iris plane. The ACD (constant) can be measured from the A-constant:

$$ACD \text{ (constant)} = (0.62467 \times A) - 68.747$$

The height of the corneal dome (H) is the distance between the cornea and the iris-plane and is calculated mathematically within the formula.

The SRK/T formula uses the same A-constant that has been originally designed for the SRK linear equation. The A-constant encompasses multiple variables that include the implant manufacturer, implant style, surgeon's technique, implant placement within the eye, and measuring equipment.

The most commonly used A-constants are as follows:

Anterior chamber lens	115.0 to 115.3
Iris-supported lens	115.5 to 115.7
Posterior chamber lens in the sulcus	
Plano-convex	115.9 to 116.2
Biconvex	116.6 to 117.2
Posterior chamber lens in the bag	
Plano-convex	117.5 to 117.8
Biconvex	117.8 to 118.8

The A-constant encompasses multiple variables that include the implant manufacturer, implant style, surgeon's technique, implant placement within the eye, and measuring equipment.

AXIAL LENGTH CORRECTION FACTOR FOR RETINAL THICKNESS

The optical axial length (LOPT) is measured by adding the retinal thickness (RETHICK) to the axial length (L):

$$LOPT = L + RETHICK \text{ and } RETHICK = 0.65696 - (0.02029 \times L)$$

THE CORNEAL REFRACTIVE INDEX

The corneal radius of curvature in millimeters is measured:

$$r = 337.5/K$$

Where K is the average keratometric reading.

Hoffer Q Formula

The Hoffer Q formula[4] uses the basic Hoffer's modification of Colenbrander's formula with a new ACD prediction formula. Hoffer studied the relationship between the ACD and the AL and found it to be a tangent curve instead of a straight line. He then tried many mathematical formula variations using the AL and average K readings, until a formula produced the desired curve. The formula consists of:
- A personalized ACD value developed from any series of one IOL style.
- A factor that increased the ACD value with increasing axial length.
- A factor that increased the ACD value with increasing corneal curvature.
- A factor that moderated the change in ACD value for extremely long (over 26 mm) and short (less than 22 mm) eyes.
- A constant added to the ACD.

The commonly used ACD values for the different IOL styles are as follows:

Anterior chamber lens	2.8 to 3.1 mm
Iris-supported lens	3.3 to 3.5 mm
Posterior chamber lens in the sulcus	
Plano-convex	3.7 to 4.0 mm
Biconvex	3.8 to 4.1 mm
Posterior chamber lens in the bag	
Plano-convex	4.3 to 4.5 mm
Biconvex	4.8 to 5.1 mm

Clinical Application

An error of 1 mm in the ACD value affects the postoperative refraction by approximately 1.0 D in a myopic eye, 1.5 D in an emmetropic eye, and up to 2.5 D in a hyperopic eye.

Figure 3-2. Calculated IOL power by the modified formulas (shaded areas) compared to the calculated powers by the SRK and Colenbrander's formulas.

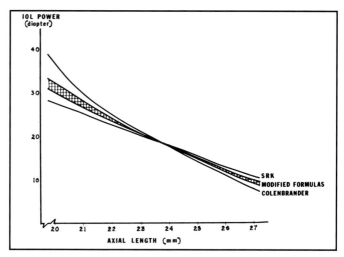

TABLE 3-1.			
IOL Power for Emmetropia			
Axial length (mm)	21.0	23.5	26.0
K readings (D)	47.0	43.5	41.0
Shammas	23.7	19.3	15.4
SRK/T	23.8	19.3	15.4
Holladay	23.7	19.3	15.1
Hoffer Q	23.9	19.3	15.3
Binkhorst	23.7	19.3	15.3

Comparing the Formulas

Figure 3-2 shows that the results of the modern formulas fall in between the results of the original theoretical (Colenbrander) and regression (SRK) formulas.

Table 3-1 shows that all modern modified formulas yield approximately the same results in an average eye (AL of 23.5 mm and K readings of 43.5 D), in a short eye with a steep cornea (AL of 21.0 mm and K readings of 47.0 D), and in a long eye with a flat cornea (AL of 26.0 mm and K readings of 41.0 D). Calculations are for a plano-convex PMMA posterior chamber IOL placed in the capsular bag. The Holladay 2 formula was not used in these comparisons.

Variations between the modern formulas still exist, especially in the short and long eyes. These variations seem to be related to changes in the corneal curvature. Figure 3-3 shows that in a 21 mm eye, the formulas differ by almost 2 D in the uncommon short eyes with flat corneas (41 D) and only by 0.2 D in the most commonly encountered short eyes with steeper corneas (47 D).

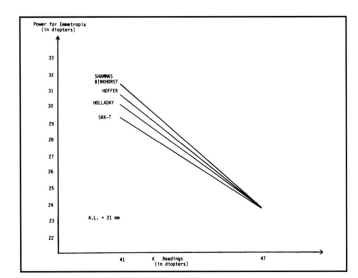

Figure 3-3. Variations in the calculated IOL power for emmetropia with the corneal curvature in the short 21 mm eye.

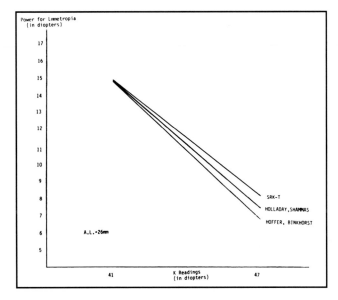

Figure 3-4. Variations in the calculated IOL power for emmetropia with the corneal curvature in the long 26 mm eye.

Similarly, in the 26 mm long eye (Figure 3-4), the formulas differ by almost 1.3 D in the unusual long eyes with steep corneas (47 D), and differ by less than 0.5 D in the most common long eyes with a flat cornea (41 D).

Clinically, these formulas have been compared in different series.[3-6] The following observations are noted:

1. All modern formulas have been found superior to the original theoretical (Colenbrander and Binkhorst) and regression formulas (SRK and SRK II).
2. There is essentially no clinically significant difference between the Holladay, SRK/T, and Hoffer Q formulas in the average eye.
3. The Holladay 2 formula improved accuracy in the short eyes with a lower mean absolute error than in the regular Holladay formula.

The Anterior Chamber Constants

Each IOL power formula uses certain constants that are specific to the said formula. The Hoffer Q formula uses the ACD constant, which is the estimated postoperative ACD. The Holladay and Holladay 2 formulas use the S factor, which is the difference between the estimated ACD and the calculated distance between the anterior corneal surface and the iris plane. The SRK/T formula uses the A-constant.

INDIVIDUALIZING THE CONSTANT

All three constants (ACD, A, or S) have to be personalized to accommodate any consistent shift that might affect the IOL power calculations, namely the surgical technique, the biometer used, and the implant's style and manufacturer. The constant has to be back-calculated for a series of 20 or more cases. It is important for the series under review to have the same parameters:

1. Same surgeon.
2. Same surgical technique:
   - Limbal, scleral, or corneal incision.
   - Planned extracapsular or phacoemulsification.
   - Capsulorrhexis or can-opener anterior capsulotomy.
   - Implant fixation in the sulcus or in the bag.
   - Wound closure with radial, horizontal, or with no sutures.
   - Postoperative steroid regimen.
3. Same biometry unit and same technician taking the measurement.
4. Same keratometer and same technician taking the readings.
5. Same IOL style from the same manufacturer.

> All constants have to be personalized for optimal results.

When the surgeon first uses a new IOL, it is best to use the designated constant, either from the manufacturer or from available tables. After having used 20 or more implants, the surgeon should personalize the constant by back-calculating the constant for each case. It is best for the initial implants to be used in average eyes between 23 and 24 mm to obtain only minimal variation in the personalized constant.

The following information is needed:
- Patient's name and number, for identification purposes.
- Preoperative AL in millimeters.
- Preoperative average K readings in diopters or radius of curvature in millimeters.
- The calculated IOL power with the formula of choice.
- The IOL used and its power.
- The expected postoperative refraction with the used IOL.
- The actual postoperative refraction. It is best to use only cases with a stable refraction and good visual acuity to avoid errors inherent to concomitant intraocular diseases.

The constant (ACD, A, or S) is then back-calculated for each case using available computer programs and calculators. The constant values are averaged to obtain the personalized constant.

CHANGING THE CONSTANT WHEN THE SURGICAL PLAN CHANGES

All constants are calculated for a posterior chamber IOL placed in the capsular bag. When circumstances dictate a change in the surgical plan, the used constant has to be modified to accommodate the situation.

All constants are calculated for a posterior chamber IOL placed in the capsular bag. When circumstances dictate a change in the surgical plan, the used constant has to be modified to accommodate the situation.

Clinical Application

Surgery is planned and IOL calculations call for a 22 D posterior chamber implant for emmetropia. During surgery, the capsule breaks and vitreous is lost. The surgeon now has to insert an anterior chamber lens and no calculations for this implant have been made. A decision on the implant power has to be made on the spot. This is where the A-constant comes in handy, no matter what formula has been used.

A-constant of the posterior chamber lens = 118.3
A-constant of the anterior chamber lens = 115.3

In most cases, the power of the IOL lens for emmetropia varies in a 1:1 relationship with the A-constant; if A decreases by 1, P decreases by 1 D also. This straight relationship adds to the simplicity and popularity of the A-constant.

In our case, the A-constant is lower by 3 if an AC lens is used; the implant power has then to be decreased by the same amount to avoid any unwanted postoperative myopia, and a 19 D AC lens should be inserted to keep the eye in the emmetropic range.

CONVERSION EQUATIONS BETWEEN THE CONSTANTS

Conversion equations are available to convert an A-constant to an ACD value or an S factor.[7] The equivalent ACD for a given A-constant can be determined from the following formula:

$$ACD = (A\text{-constant} \times 0.5836) - 63.896$$

The equivalent S constant for a given A-constant can be determined from the following formula:

$$S = (A\text{-constant} \times 0.5663) - 65.60$$

			TABLE 3-2.		
		Lens Constant Conversion Table			
A-constant	ACD value	S factor	A-constant	ACD value	S factor
115.0	3.21	-0.48	117.0	4.38	0.66
115.2	3.33	-0.36	117.2	4.50	0.77
115.4	3.45	-0.25	117.4	4.62	0.88
115.6	3.56	-0.14	117.6	4.73	1.00
115.8	3.68	-0.02	117.8	4.85	1.11
116.0	3.80	0.03	118.0	4.97	1.22
116.2	3.91	0.20	118.2	5.08	1.34
116.4	4.03	0.32	118.4	5.20	1.45
116.6	4.15	0.43	118.6	5.32	1.56
116.8	4.27	0.54	118.8	5.43	1.68

Clinical Application

An acrylic IOL placed in the capsular bag has an A-constant of 118.4. The corresponding ACD value and S factor can be calculated:

$$ACD \text{ value} = (118.4 \times 0.5836) - 63.896 = 5.20$$
$$S \text{ factor} = (118.4 \times 0.5663) - 65.60 = 1.45$$

Table 3-2 gives the conversion values of the most commonly used lens constants.[7]

References

1. Holladay JT, Prager TC, Chandler TY, Musgrove KH. A three-part system for refining intraocular lens power calculations. *J Cataract Refract Surg.* 1988;14:17-24.
2. Holladay JT. *Holladay IOL Consultant User's Guide and Reference Manual.* Houston, TX: Holladay LASIK Institute; 1999.
3. Retzlaff J, Sanders DR, Kraff MC. Development of the SRK/T intraocular lens implant-power calculation formula. *J Cataract Refract Surg.* 1990;16:333-340. Errata 1990;16:528.
4. Hoffer KJ. The Hoffer Q formula: A comparison of theoretic and regression formulas. *J Cataract Refract Surg.* 1993;19:700-712. Errata 1994;20:677.
5. Shammas HJF. The fudged formula for intraocular lens power calculations. *Am Intraocular Implant Soc J.* 1982;8:350-352.
6. Hoffer KJ. Clinical results using the Holladay 2 intraocular lens power formula. *J Cataract Refract Surg.* 2000;26:1233-1237.
7. Holladay JT. International intraocular lens & implant registry. *J Cataract Refract Surg.* 2003;29:176-197.

The Olsen Formula

Thomas Olsen, MD

The early intraocular lens (IOL) power calculation formulas from the 1970s[1-3] were "theoretic" formulas in the sense that they used optical formulas in the calculation of lens power. The clinical accuracy with these formulas was disappointing, however, and a better accuracy was reported using linear regression formulas on large empirical datasets with minimal optical foundation.[4,5]

The Olsen formula was developed in the late 1980s[6-11] at a time when the regression formulas were dominant and the optical formulas were ill regarded. It was the clinical impression that much room was still left for improvement, and the Olsen formula was developed as a result of a critical analysis of the conditions for accurate IOL calculation. The aim was to develop a formalism using as few assumptions as possible within the realm of Gaussian Optics and to make this model applicable to the optics of the pseudophakic eye. The Olsen formula addresses four areas of concern: the calculation of the corneal power, the measurement of the axial length (AL), the anterior chamber depth (ACD) prediction, and the IOL optics.

Calculation of the Corneal Power

One of the first problems in the calculation of IOL power is how to calculate the power of the cornea. Because conventional keratometers only measure the curvature of the front surface of the cornea and regard this as the only refracting surface (Table 4-1, equation 1), the curvature of the posterior surface needs to be assumed. It appears that the fictitious "refractive index" still in use by most keratometers (1.3375) is not based on physiological considerations, but seems to have been chosen solely because of its nice numbers: given a curvature of 7.5 mm, the corneal power comes out at 45.0 D on the scale of the keratometer.

However, if one assumes the curvature of the posterior surface of the cornea to be in proportion to the anterior surface, as in the exact schematic eye of Gullstrand (cur-

---

TABLE 4-1.
Formulas for Power Calculation of Two Refracting Systems

The simplified calculation of corneal power is similar to the formula describing the dioptric power of a single refracting spherical surface:

$$D = (n2 - n1) \: / \: r \qquad (1)$$

Where D = power in diopters, n2 = refractive index of the second medium (cornea), n1 = refractive index of first medium (air = 1.0), and r = radius of curvature in m.

The total power of two refracting systems in combination is given by:

$$D12 = D1 + D2 - t \times D1 \times D2 \qquad (2)$$

Where D12 = total dioptric power, D1 = power of first system, D2 = power of second system, t = reduced distance in m between the two refracting systems. The reduced thickness is defined as:

$$t = T/n \qquad (2a)$$

where T = real thickness in m and n = refractive index.

The principal planes are given by:

$$p1 = D2 \times t/D12 \qquad (2b)$$
$$p2 = -D1 \times t/D12 \qquad (2c)$$

---

vature ratio 6.8 mm/7.7 mm = 0.883), and uses the true, physiological refractive index of 1.376, the equivalent "refractive index" can be found to be 1.3315.[6] This index is lower than the index of 1.3333 used in the Binkhorst formula and lower than most formulas using the dioptric value of a 1.3375–calibrated conventional keratometer.[4,12,13]

The value of 4/3 (1.3333) used by Binkhorst was chosen not because of its superior physiological value, but because it offered a correction to the flattening of the cornea observed after cataract surgery with the technique used in those days when large incisions were the standard. This assumption does not apply to the current phacoemulsification technique with its small incision, which has minimal effect on the average corneal power.

The difference in calculated corneal power using 1.3315 in the Olsen formula versus 1.3375 with the conventional keratometer amounts to almost 1 D for a normal cornea, and might therefore introduce a significant *a priori* error in the calculation of IOL power. What evidence do we have, then, to decide on the most accurate refractive index?

The posterior curvature of the cornea has recently been studied by Dunne, Royston and Barnes,[14] who photographed the Purkinje II images from the corneal endothelium to derive normal values as well as the variability of the posterior corneal curvature. Their results revealed a steeper curvature of the posterior surface than previously assumed by Gullstrand and others, showing a normal value for the ratio between anterior and posterior surface of 0.823. The value of 0.823 is the equivalent of a "refractive index" of only 1.3283, which seems to be the lowest value currently proposed. The difference in the calculated power between 1.3315 (Olsen) and 1.3283 (Dunne et al[14]) amounts to about 0.4 D in the average case.

Another lesson on the effective corneal power can be learned from the power determination of an IOL *in situ*.[15] It is possible to use an extended-range keratometer to measure the apparent front curvature of a lens within the eye using the Purkinje-Sanson III images arising from its front surface. When transforming the apparent curvature to the true curvature, one needs to correct for the effective power of the cornea (and measure the ACD) in order to have an accurate estimate. It was found that accurate estimations[6] of the front lens surface could be obtained with no significant off-set errors using the corneal refractive index of 1.3315. If the assumed index of refraction was higher than this (eg, 1.3333), a significant off-set error and ultimately an error in the IOL power calculation would be the result. These observations therefore support the view that the physiological refractive index of the cornea is in the lower end of that normally assumed (and lower than in most IOL power formulas).

Measurement of the Axial Length

The AL measured by ultrasound need not be the true AL in the optical sense. Firstly, the "retinal" spike seen on ultrasound originates from the vitreoretinal interface and there is a small distance from here to the sensory outer segments. Secondly, when A-scan measurements are taken as applanatory readings (contact technique), there is further *a priori* reason to believe that the ultrasonic measurements are too short due to compression of the cornea. So the term *retinal thickness* was introduced in the early theoretic formulas (and used by the Olsen formula) as a corrective term to be added to the ultrasonic readings in order to eliminate a systematical off-set error in the IOL power predictions.

It is an old observation that the large errors in IOL calculation are more likely to arise in extreme short and long eyes. In the early days of IOL power calculation, this AL bias was considered more likely to occur in the theoretic formulas than in the regression formulas. This was later shown to be due to an inaccurate ACD prediction. Before blaming the formula, however, it might be necessary to consider the velocity assumptions of the AL measurements by ultrasound.

Normally, the average velocity of 1550 meters per second (m/s) is assumed for the distance from the corneal surface to the retina. This is based on the average values for an average phakic eye for which 1532 m/s is assumed to be the velocity for the anterior chamber and the vitreous and 1640 m/s is assumed to be the velocity for the biological lens.[16] In case of extreme myopia or hyperopia, the average velocity changes. In the very short eye, the lens makes up a larger part of the whole distance and the average velocity needs to be increased and vice versa for a long eye. If the lens thickness was the same, this would be easy to correct, but this is not the case, as the lens tends to flatten in a long eye and thicken in a short eye.

Therefore, to correct the AL according to the shifts in velocity, one must take into account the actual statistical relationship between the individual ultrasonic distances. Based on a large series with individual measurements of the lens thickness, anterior chamber and the vitreous distance, the following regression formula was found to give the lens thickness (in mm) as a function of the AL:

$$LThick = 6.44 - 0.082 \times Ax$$

Using this relation, the AL can be corrected according to the equation:

$$RealAx = (Ax/MeanVel - LThick/LensVel) \times AqueousVel + LThick$$

Where RealAx is the corrected axial length, Ax is the axial length measured assuming an average velocity of MeanVel (= 1550 m/s), LensVel is the average velocity of the biological lens (= 1640 m/s) and AqueousVel is the average velocity of the aqueous and the vitreous (= 1532 m/s).

With the recent introduction of optical partial coherence interferometry (PCI) for the measurement of AL[17] (Zeiss IOLMaster), we may now have a more "clean" measurement of the optical path so that we may avoid the addition of a retinal thickness corrective term. Recent investigations seem to confirm that when the AL is measured by the PCI technique and used in the Olsen formula, the need for a corrective retinal thickness term is minimized and may be eliminated (see clinical study in Table 4-3).

The ACD Prediction

One of the major drawbacks of the earlier theoretic formulas was the lack of empirical data on the postoperative position of the implant (the postoperative ACD). The early Binkhorst formula (Binkhorst I) used a constant value for the ACD, and was later replaced by Binkhorst II, in which the postoperative ACD was dependent on the AL according to the simple formula:

$$ACDpost = ACDmean \times Ax/23.45$$

Where ACDmean is the average ACD (so-called ACD constant) of a given IOL type and Ax is the axial length in millimeters (Ax < 26 mm).

The dependence of the ACD on the AL is important in order to avoid bias in the calculation of the IOL power: if not corrected for the fact that a long eye has a deep postoperative ACD, the ACD is predicted too deep in short eyes and too shallow in long eyes. The result is a myopic error (overestimated IOL power) in short eyes and a hyperopic error (underestimated IOL power) in long eyes. To illustrate the problem, one might ask the following question: if the postoperative ACD were to be predicted with ultimate accuracy, do we still need a corrective term?

This question was studied by measuring the actual postoperative ACD in a large series (about 1000 cases), taking the individual postoperatively measured ACD and substituting the predicted postoperative ACD with the true, postoperative ACD in the calculations.[18] This experiment gave the conclusion that no off-set errors were observed over the entire AL range! A very important conclusion to be drawn from this study was that we did not need any fudge factors (A-constants and other empirically defined constants) to adjust the IOL power calculations, all we needed was an accurate prediction of the physical position of the IOL and we had a home run.

So, the task was to describe methods by which the postoperative ACD could be predicted in a given eye based on the preoperative measurements of that eye.[19] In the hands of the author, one of the best methods was to perform regression analysis on the actual postoperative ACD, as measured by ultrasound, and to study its dependence on several preoperatively defined variables. These variables include the following axial distances: the corneal height, the anterior chamber depth, the lens thickness, and the axial length, making a formula of the form:

$$ACDpost = a + b1 \times H + b2 \times ACDpre + b3 \times Lthick + b4 \times Ax$$

Where H is the corneal height; ACDpre is the anterior chamber depth; Lthick is the lens thickness; Ax is the axial length; b1, b2, b3, and b4 are regression coefficients; and a is an off-set constant that can be derived from the average values. The current regression coefficients (2003) are as follows:

ACDpost = ACDMean + 0.12 x H + 0.33 x ACDpre + 0.30 x Lthick + 0.10 x Ax–5.18

Where ACDMean is the average ACD as measured by ultrasound in a representative sample of cases. This formula applies to phakic eyes with normal anterior segment anatomy only. The coefficients will change in cases of pseudophakic or aphakic eyes.

ACD prediction takes into account the corneal height, the preoperative ACD, the lens thickness, and the axial length.

The corneal height is defined as the height of the spherical segment made by the corneal sphere from the limbal plane. It can be calculated from the corneal curvature and the corneal diameter, the latter by taking an average value or refined using the actual white-to-white distance of the cornea. The corneal height was first used by Fyodorov in the power calculation for an iris clip IOL,[3] and was later adopted by Olsen for anterior chamber lenses.[6,7] For posterior chamber IOLs, the significance of the corneal height seemed at first to be of minor importance as compared to the AL, the preoperative ACD, and the lens thickness, in that order. Later it was reintroduced by Holladay in the calculation of the so-called surgeon factor, which was defined as the distance from the corneal height to the effective optical plane of the IOL.[12] Today, as most surgeons prefer to do a continuous curvilinear capsulorrhexis and place the IOL in the capsular bag, there seems to be less room for surgical variability and much room for good ACD prediction.[10]

With current improvements in AL measuring techniques and overall surgical standardization, the ongoing demands on more accurate IOL power predictions may call for the most extensive ACD methods utilizing all known predictors. Today, most of the newer generation theoretic formulas recognize the importance of using more than just the AL in predicting the ACD[20] (Table 4-2).

What is the significance, then, of an accurate ACD prediction as compared to the other errors in IOL power calculations?

To answer this, a study was performed to quantify and split the total error of IOL calculations into three components: errors arising from the axial length measurements, errors arising from the corneal power measurements, and errors arising from the ACD prediction.[21] The results from this study, which involved complete pre- and postoperative biometry of more than 500 patients, showed that 54% of the error could be contributed to AL errors, 8% to corneal power errors, and 38% to errors in the ACD estimation, when using a fixed ACD in the calculations (eg, the Binkhorst I formula). However, with optimal prediction of the ACD at that time,[21] the error originating from the ACD prediction decreased by about 50%, thereby decreasing the total error by about 20%.

The conclusion from this study was that the ACD prediction does play a significant role in the IOL power calculation and makes up the heart of the IOL formula, provided other errors have been accounted for. If the error of the AL measurement were to be reduced, the relative importance of the ACD prediction would of course be even greater.

TABLE 4-2.					
Preoperative Distances Used in the					
Individual Prediction of the Postoperative ACD					
Formula	**Year**	**Axlen**	**Cheight**	**ACDpre**	**Lthick**
Binkhorst II	< 1980	Yes	No	No	No
SRK I + II	1981, 1988	Yes	No	No	No
SRK/T	1990	Yes	Yes	No	No
Olsen	1987 - 1995	Yes	Yes	Yes	Yes
Holladay 1	1988	Yes	Yes	No	No
Holladay 2	1996	Yes	Yes	Yes	Yes
Haigis	1996	Yes	Yes	Yes	Yes

Axlen = axial length, Cheight= corneal height, ACDpre = preoperative anterior chamber depth, Lthick = lens thickness.

The IOL Optic

Apart from knowing the position of the implant, in order to calculate the IOL power according to Gaussian Optics, it is necessary to know the position of the principal planes of the IOL optic. Most IOL manufacturing companies are willing to list the curvatures of the front and back surface of the IOL along with other necessary physical information and this makes it possible to calculate the position of the principal planes for a given power. The current ANSII standardization requires the dioptric power of an IOL to be labeled as the effective power when embedded in aqueous (refractive index 1.336). Knowing the refractive index as well as the curvature of the anterior and posterior surface of the IOL optic and its central thickness, the equivalent power of the implant and its principal planes can be calculated according to well-known formulas (see equations 1 and 2 in Table 4-1).

The position of the principal planes is important in determining the effective power of the lens within the eye. For instance, if the power for a PMMA biconvex IOL with a 1:2 optic configuration (meaning that the front/back curvature has the ratio 1/2) has been determined to be 22 D, and this IOL is to be exchanged for a lens of the same material and at the same position within the eye but with a reverse 2:1 configuration, then this lens would have to be about 0.5 D stronger to achieve the same refractive result. This is the reason for shift in the SRK A-constant when the optic configuration of a given IOL is changed but the intended placement (ie, the capsular bag) is unchanged.

Optimization of the Formula

Although the accuracy of the measuring equipment has improved much over the last decades, differences in calibration, differences in the techniques for AL determination (immersion versus contact technique[22]) as well as surgical variation, may account for average differences in the postoperative position of the IOL. These systematical differences call for the use of a personalized constant in the IOL calculation, as can be studied from the follow-up of surgical cases.

The concept of personalized surgical constants for a given IOL type was elaborated by the regression formulas and widely known from the SRK A-constant. The SRK A-constant can be deduced mathematically from the average power needed for emmetropia for the eye with an average AL and average keratometry. Besides information of the postoperative position of the IOL, the A-constant includes all systematical errors arising in the clinical setting, including IOL optic configuration, measuring errors, differences in surgical technique, as well as errors in the optical calculation. It should be noted, however, that in order for surgeon A to be able to use the A-constant derived by surgeon B, surgeon A should have an identical measuring apparatus in addition to identical placement of the IOL within the capsular bag (we assume for simplicity, of course, that both surgeons do not change the corneal power in the average case). If these conditions are not met, surgeon A needs to do his own follow-up study on a sufficient number of cases.

Because the A-constant is the sum of several sources of error, the number of cases needed for an accurate estimate is considerable. It is the experience of this author that the standard deviation of the A-constant in a typical sample of cases is around ± 1 D. If the desired accuracy of the A-constant were to be within ± 0.25 D (95% confidence limit = ± 2 Standard Errors of the Mean, [SEM]), the number of observations "n" can be calculated from the formula:

$$\pm\ 0.25\,D = \pm\ 1.0\,D \times 2\ /\ sqrt(n)\ \Rightarrow n = 64$$

With the Olsen formula or other formulas based on exact ACD predictions, all that is needed is knowledge of the average ACD for a given IOL type. It is the experience of this author that the standard deviation of the postoperative ACD in a typical sample of cases is around ± 0.25 mm. As this corresponds to an IOL power error of about ± 0.5 D, the number of observations needed to obtain an ACD constant with an equivalent ± 0.25 D accuracy (95% limits) can be found from the formula:

$$\pm\ 0.25\,D = \pm\ 0.50\,D \times 2\ /\ sqrt(n) \Rightarrow n = 16$$

Hence, the number of cases needed for an optimized ACD-based constant is only one fourth of that needed from a refraction-based constant.

When the actual postoperative ACD is known in a given eye, it is easy to do a backward IOL calculation and calculate the effective AL from the known refraction, corneal power, and IOL power.[23] By taking the difference between this *a posteriori* calculated AL and the preoperatively measured AL, one can derive the retinal thickness factor needed in the IOL calculation. This factor is a true constant depending only on the measuring apparatus and need not be changed when changing the IOL type.

Formula Validation

Although accurate IOL power calculations can be achieved using clinical measurements with minimal corrections, the use of Gaussian Optics may have its limitations. Newer wavefront techniques have broadened our understanding of ocular imagery and may give a better description of the total optical system of the eye. The assumption that the effective "focus" is a paraxial ray focus may not hold when spherical aberrations and other higher order aberrations are taken into account.

However, before methods from optical physics are implemented in the IOL power calculations, it has to be remembered that the retina may play a role of its own in selecting the effective focus. One such factor is the Stiles-Crawford effect, which predicts the sensory effect of a peripheral skew ray to be only a fraction of that of a central paraxial ray, thereby compensating somewhat for the effect of spherical aberration,[24] and justifying the application of paraxial imagery in the field of IOL power calculation.

Any formula claiming accuracy over current formulas should document not only the overall accuracy on an optimized dataset, but also provide the evidence that the empirical component of the formula, ie, the ACD prediction, actually works better in a large range of series that includes long and short eyes.

CLINICAL STUDY

Ninety-eight routine cataract patients (26 men and 72 women, age range 39 to 92 years) were investigated. Biometry was performed using conventional contact A-scan technique (Tomey AL-2000 [Tomey Corporation, Nagoya, Japan]), as well as by partial optical coherence interferometry (Zeiss IOLMaster [Zeiss Humphrey Systems, Dublin, CA]). The surgical technique was phacoemulsification with implantation of a foldable, acrylic lens implant in the capsular bag after continuous curvilinear capsulorrhexis. Postoperative follow-up examinations were made 2 to 3 months after surgery with complete biometry including the measurement of postoperative ACD with ultrasound.

Four sets of formulas were evaluated: SRK I, SRK/T, Holladay, and Olsen. For the formula evaluation, each formula was optimized in retrospect by adjusting the appropriate constants. This involved the *a posteriori* calculation of the A-constant for the SRK I formula, the surgeon's factor for the Holladay formula, the measurement of the average postoperative ACD for the Olsen formula, and the estimation of the AL correction factor ("retinal thickness") for the Olsen formula. The corresponding constants are displayed in Table 4-3. Note that the A-constant is not a constant, but changes as the formula is changed from SRK I to SRK/T and changes according to the biometry technique.

The mean prediction errors are shown in Table 4-4. There was a significant improvement in accuracy from the early regression formula (SRK I) to the newer theoretic formulas (SRK/T, Holladay, Olsen). A small improvement was noted with the optical coherence interferometry (Zeiss) as compared to conventional ultrasound when using SRK/T, Holladay, and Olsen, but not with the SRK I formula. The highest accuracy was seen with the Olsen formula and the PCI technique, giving an absolute error of 0.46 D in the average case (Figure 4-1).

When the actual postoperative measurements were used in the "prediction" of IOL power, the PCI technique was by far the most accurate, giving an estimated IOL power within ±1.0 D in 96% of the cases with an average absolute error of 0.34 D (Figures 4-2 and 4-3).

CONCLUSION FROM CLINICAL STUDY

The PCI technique offered some advantage over contact A-scan ultrasound in the calculation of IOL power, which can be exploited by an accurate IOL power formula. However, the improvement in preoperative predictions is not as large as expect-

TABLE 4-3.
Lens and Formula Constants (Optimized)

Used in the comparison of the accuracy with contact A-scan ultrasound and partial optical coherence interferometry (PCI, Zeiss IOLMaster) in 98 consecutive cataract cases. The IOL was a foldable acrylic lens implant. Note the different lens constants with the SRK method and the Holladay method according to axial length measuring technique.

Formula	SRK I	SRK /T	Holladay	Olsen	Olsen
Constant	A-constant	A-constant	Surg. Factor	ACD	Retinal thickn
Value US	119.4 D	118.8 D	1.63	4.54 mm	0.26 mm
Value PCI	120.0 D	119.3 D	1.97	4.54 mm	0.02 mm

TABLE 4-4.
Formula Accuracy with Conventional Contact A-Scan Ultrasound

The error is defined as the difference between the observed and the predicted refraction. All formulas have been optimized, ie, all formula constants have been adjusted to give an average numerical prediction error of zero. N = 98.

Ultrasound	SRK I	SRK/T	Holladay	Olsen
Numerical error	0 D	0	0	0
Absolute error	0.92 D	0.53 D	0.50	0.46
Standard deviation	± 1.19 D	± 0.69 D	± 0.66 D	± 0.61 D
Range	-3.00 – +3.17D	-1.59 – +1.50 D	-1.70 – +1.48 D	-1.44 – +1.82 D
Errors > 2 D	12	0	0	0
Errors > 1 D	38	22	16	12

(continued)

Table 4-4 (Continued).
Formula Accuracy with Partial Coherence Interferometry

Formula accuracy using optical partial coherence interferometry (PCI, Zeiss IOLMaster) for the axial length measurement. The error is defined as the difference between the observed and the predicted refraction. All formulas have been optimized, ie, all formula constants have been adjusted to give an average prediction error of zero. N = 110.

PCI (Zeiss IOLMaster)	SRK I	SRK T	Holladay	Olsen
Numerical error	0 D	0 D	0 D	0 D
Absolute error	0.94 D	0.51 D	0.47 D	0.44 D
Standard deviation	± 1.20 D	± 0.64 D	± 0.60 D	± 0.58 D
Range	-3.38 – +3.19 D	-1.77 – +1.48 D	-1.84 – +1.40 D	-1.66 – +1.45 D
Errors > 2 D	13	0	0	0
Errors > 1 D	42	15	17	11

Figure 4-1. The mean absolute prediction error (diopters) according to IOL formula (optimized) and axial length technique. There was a significant improvement in accuracy from the SRK I regression formula to the newer theoretic formulas (SRK/T, Holladay, Olsen), and a small improvement with the optical coherence interferometry (Zeiss) with all formulas except with the SRK I formula.

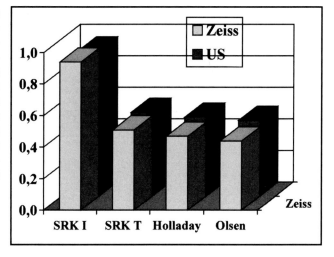

ed from the high technical precision of the technique. This result may be due to the variability of the biological lens, the (unknown) refractive index of which may add to the measurement error.

However, in the pseudophakic eye, the PCI technique offers by far the highest accuracy in the calculation of IOL power. With an average accuracy of 0.34 D, it is possible to use the PCI technique together with an accurate IOL formula (like the Olsen formula) as a diagnostic tool to measure the actual IOL power within the eye. This is important when dealing with refractive surprises caused by measurement errors, formula errors, or errors in IOL power labeling.

Figure 4-2. The refractive prediction error according to the Olsen formula and PCI technique (Zeiss IOLMaster) for the determination of the phakic axial length with no axial length corrections. The mean error was -0.01 D with a standard deviation of +0.57 D. The average absolute error on this optimized dataset was 0.43 D (n = 98).

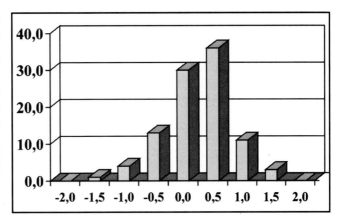

Figure 4-3. The refractive prediction error when "predicting" the actual IOL power from the postoperative measurements, using PCI for the determination of the pseudo-phakic axial length with no corrections. The mean error was 0.01 D with a standard deviation of +0.48 D. The average absolute error was 0.34 D (n = 98).

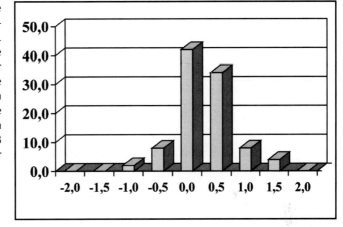

References

1. Colenbrander MC. Calculation of the power of an iris clip lens for distant vision. *Br J Ophthalmol.* 1973;57:735-740.
2. Binkhorst RD. The optical design of intraocular lens implants. *Ophthalmic Surg.* 1975; 6:17-31.
3. Fyodorov SN, Galin MA, Linksz A. Calculation of the optical power of intraocular lenses. *Invest Ophthalmol.* 1975;14:625-628.
4. Sanders D, Retzlaff J, Kraff M, Kratz R, Gills J, Levine R, Colvard M, Weisel J, Loyd T. Comparison of the accuracy of the Binkhorst, Colenbrander, and SRK implant power prediction formulas. *American Intra-Ocular Implant Society Journal.* 1981;7:337-340.
5. Thompson JT, Maumenee AE, Baker CC. A new posterior chamber intraocular lens formula for axial myopes. *Ophthalmology.* 1984;91:484-488.
6. Olsen T. On the calculation of power from curvature of the cornea. *Br J Ophthalmol.* 1986;70:152-154.
7. Olsen T. Theoretical approach to intraocular lens calculation using Gaussian optics. *J Cataract Refract Surg.* 1987;13:141-145.

8. Olsen T, Thim K, Corydon L. Theoretical versus SRK I and SRK II calculation of intraocular lens power. *J Cataract Refract Surg.* 1990;16:217-225.
9. Olsen T, Thim K, Corydon L. Accuracy of the newer generation intraocular lens power calculation formulas in long and short eyes. *J Cataract Refract Surg.* 1991;17:187-193.
10. Olsen T, Gimbel H. Phacoemulsification, capsulorrhexis, and intraocular lens power prediction accuracy. *J Cataract Refract Surg.* 1993;19:695-699.
11. Olsen T, Corydon L, Gimbel H. Intraocular lens power calculation with an improved anterior chamber depth prediction algorithm. *J Cataract Refract Surg.* 1995;21:313-319.
12. Holladay JT, Prager TC, Chandler TY, Musgrove KH, Lewis JW, Ruiz RS. A three-part system for refining intraocular lens power calculations. *J Cataract Refract Surg.* 1988;14:17-24.
13. Sanders DR, Retzlaff JA, Kraff MC, Gimbel HV, Raanan MG. Comparison of the SRK/T formula and other theoretical and regression formulas. *J Cataract Refract Surg.* 1990;16: 341-346.
14. Dunne MC, Royston JM, Barnes DA. Normal variations of the posterior corneal surface. *Acta Ophthalmol (Copenh).* 1992;70:255-261.
15. Olsen T. Measuring the power of an in situ intraocular lens with the keratometer. *J Cataract Refract Surg.* 1988;14:64-67.
16. Jansson F, Kock E. Determination of the velocity of ultrasound in the human lens and vitreous. *Acta Ophthalmol (Copenh).* 1962;40:420-433.
17. Drexler W, Findl O, Menapace R, Rainer G, Vass C, Hitzenberger CK, Fercher AF. Partial coherence interferometry: A novel approach to biometry in cataract surgery. *Am J Ophthalmol.* 1998;126:524-534.
18. Olsen T, Corydon L. We don't need fudge factors in IOL power calculation. *Eur J Implant Refract Surg.* 1993;5:51-54.
19. Olsen T. Prediction of intraocular lens position after cataract extraction. *J Cataract Refract Surg.* 1986;12:376-379.
20. Haigis W, Lege B, Miller N, Schneider B. Comparison of immersion ultrasound biometry and partial coherence interferometry for intraocular lens calculation according to Haigis. *Graefes Arch Clin Exp Ophthalmol.* 2000;238:765-773.
21. Olsen T. Sources of error in intraocular lens power calculation. *J Cataract Refract Surg.* 1992;18:125-129.
22. Olsen T, Nielsen PJ. Immersion versus contact technique in the measurement of axial length by ultrasound. *Acta Ophthalmol (Copenh).* 1989;67:101-102.
23. Olsen T. Calculating axial length in the aphakic and the pseudophakic eye. *J Cataract Refract Surg.* 1988;14:413-416.
24. Olsen T. On the Stiles-Crawford effect and ocular imagery. *Acta Ophthalmol (Copenh).* 1993;71:85-88.

The Haigis Formula

Wolfgang Haigis, PhD

The Thin Lens Formula

Popular formulas for intraocular lens (IOL) power calculation like the Hoffer Q,[1] the Holladay-1,[2] and the SRK/T[3] are based on the thin lens optics, with the exception of the empirical SRK I/II formulas.[4,5] In thin lens optics, the cornea and the lens (crystalline or IOL) are replaced by infinitely thin lenses (Figure 5-1) with two refractive powers—D_C (corneal power) and D_L (IOL power)—separated by a distance d. This fictitious distance is sometimes referred to as *optical* anterior chamber depth (ACD), which has no measurable counterpart, in contrast to the acoustic ACD measured by ultrasound. Holladay[6] uses the term *effective lens position* (ELP) for d.

Thus, all "theoretical" formulas may be reduced to the elementary thin lens formula:

$$D_L = \frac{n}{L-d} - \frac{n}{\frac{n}{z} - d} \quad \text{with } z = D_C + \frac{R_x}{1 - R_x \times d_x} \quad \text{and} \quad D_C = \frac{n_C - 1}{R} \quad (1)$$

Where L is the axial length, R is the corneal radius of curvature, R_x is the refraction, n=1.336, and d_x is vertex distance (=12 mm).

The mentioned theoretical formulas differ in how measured values are translated into the variables L, d and D_C of equation 1. Table 5-1 gives an overview of how different formulas handle this conversion compared to the Haigis formula.[7-11]

The main differences between theoretical formulas lie in the prediction functions for the optical ACD d, ie, in the terms $d(Holladay)$, $d(SRKT)$, $d(Hoffer)$, etc. These values vary with the axial length (AL) and consist of individual constants like the A-constant, the surgeon factor (sf), or Hoffer's personalized ACD (pACD, see Table 5-1). All of these constants may readily be transformed into each other.[12,13] For exam-

Figure 5-1. Thin lens model of an emmetropic eye (ie, additional spectacle lens for $R_x=0$ [ametropia] omitted): cornea and lens are reduced to infinitely thin lenses.

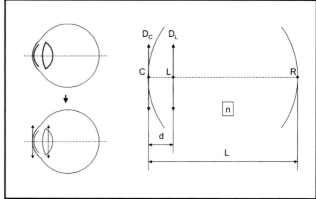

TABLE 5-1.
Differences in Theoretical IOL Formulas

Formula	n_C	L	d	IOL const
SRK/T	1.3330	AL + f(AL)	d(SRK/T)	A-const
Holladay 1	4/3	AL + 0.2	d(Holladay)	sf
Hoffer Q	1.3360	AL	d(Hoffer)	pACD
Haigis	1.3315	AL	d(Haigis)	a_0, a_1, a_2

All are based on thin lens optics (Equation 1).

ple, if A-constant = 118.0, then sf = 1.223, and pACD = 4.97. Figure 5-2 shows such prediction curves (optical ACD d versus axial length) for the Holladay 1, SRK/T, and Hoffer Q formulas, all based on an A-constant 118.0.

Since all IOL constants may be calculated from each other, there is basically just one constant, ie, one number characterizing a given lens for all available powers, regardless of shape factor, lens material, diameter, etc. This, in the author's opinion, is insufficient for a meaningful lens characterization, as will be illustrated below.

Effect of Lens Geometry on IOL Position

Following the concept of Norrby[14] and taking the capsular bag equator EP as a measure for the IOL position and considering small, medium, and long eyes, the schematic AL dependence of Figure 5-3 is obtained. Small eyes have a shallower ACD with the capsular bag equator lying more anteriorly; in long eyes the lens lies deeper in the eye with the bag equator positioned more posteriorly.

This behavior is backed up by clinical findings on 15123 eyes,[15] yet mostly unpublished data (Figure 5-4). From preoperative high precision immersion ultrasound measurements of anterior chamber depth (AC) and lens thickness (LT) as shown, the AL dependence of EP was deduced under the assumption EP=AC + 0.4 LT.

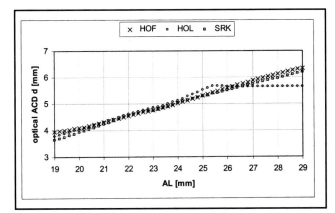

Figure 5-2. Prediction curves for the optical ACD d in equation 1 for different theoretical formulas and an A-constant of 118.0 (for SRK/T [SRK]), equivalent to sf=1.223 (for Holladay 1 [HOL]), pACD=4.97 (for Hoffer Q[HOF]).

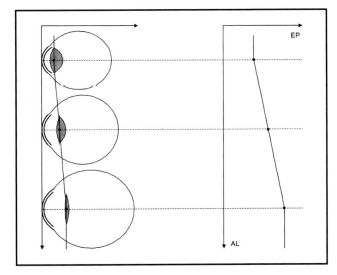

Figure 5-3. Schematic axial length dependence of the position EP of the capsular bag equator.

Figure 5-5 gives a schematic representation of the positions of the image principal planes of lenses with different shape factors and geometry (here: plano-convex and asymmetric biconvex) in eyes with different ALs. It is this position (of the image principal plane) which essentially determines d in Equation 1. It is clearly evident from Figure 5-5 that different lenses are characterized by different AL dependences of their optical ACDs.

Thus, a curve (eg, prediction of d vs AL) rather than a number (IOL constant) seems more apt for the characterization of an IOL.

Calculation According to Haigis

Using the thick lens algorithm[16] for IOL calculation in the 80s, we, like others,[17-19] were looking for ways to predict the postoperative IOL position by means of multiple regression analysis performed on preoperative data.[20] We found the main con-

Figure 5-4. Anterior chamber (AC), lens thickness (LT), and assumed position of capsular bag equator (EP) vs. axial length for n=15123 eyes. Data points: running means; assumption for EP: EP=AC+0.4 LT.

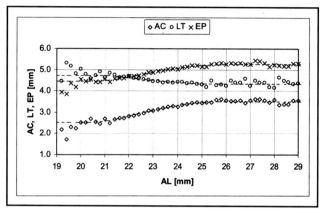

Figure 5-5. Schematic representation of lenses of different shape factors implanted into eyes with different axial lengths. The gray lines near the anterior vertex of the plano-convex lens and the posterior vertex of the biconvex lens denote the positions of the image principal planes of the two lens types.

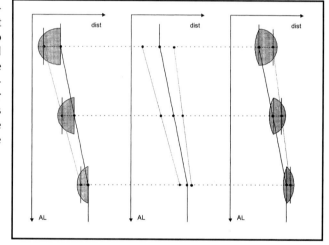

tributions to the predictability of postoperative AC (AC_{post}) to stem from the AL and the preoperative AC (Table 5-2). Therefore, we predicted the (acoustically measurable) postoperative anterior chamber depth AC_{post} according to:

$$AC_{post} = c_0 + c_1 \, AC + c_2 \, AL$$

The constants c_0, c_1, and c_2 followed from a double linear regression analysis. Since the thick lens formula requires lens design data (eg, radii of curvature and center thicknesses for all lens powers, precise refractive indices, etc.) that manufacturers were hesitant to release, we turned back again to the thin lens formula (equation 1). This time, however, we applied the regression prediction to the optical ACD, ie,[7]

$$d(\text{Haigis}) = d = a_0 + a_1 \, AC + a_2 \, AL \qquad (2)$$

The constants a_0, a_1, and a_2 were found to be quite typical for a given IOL.[21] This led to the idea of using this set of numbers for the characterization of different IOLs.

Haigis formula uses three (a_0, a_1, a_2) constants for a given IOL instead of one.

TABLE 5-2.

Correlation Coefficient for the Prediction of the Postoperative Anterior Chamber Depth

Parameter Used For Regression	Correlation Coefficient [%]
AC	68
LT	36
AL	44
CC	6
AC, LT	68
AC, AL	70
AC, LT, AL	70
AC, LT, AL, CC	71

Correlation coefficient for the prediction of the (acoustically measurable) postoperative anterior chamber depth $AC_{post} = a_0 + a_1 AC + a_2 LT + a_3 AL + a_4 CC$ for plano-convex lenses type CILCO KR2U, if different parameters or combinations thereof are used. AC= preoperative anterior chamber depth; LT= lens thickness; AL= axial length; CC= corneal curvature (radius).[20]

Olsen[18,19] uses a similar regression approach with even more variables to predict postoperative IOL positions. However, apart from being characterized by their classical ACD constants, no further differentiation is made between different lenses. Likewise, Holladay's IOL calculation program (*Holladay IOL Consultant*, Holladay LASIK Institute, Houston, TX) does not use—to the best of the author's knowledge—more than one lens constant to characterize a given IOL.

An essential aspect of equation 2 lies in the fact that with three constants (a_0, a_1, and a_2) it is possible to model the AL dependence of the optical ACD of a given lens, thus characterizing the IOL by a curve rather than a number directly derived from equation 2. Since the preoperative AC itself is dependent on AL (see Figure 5-4), *d(Haigis)* as defined by equation 2 is a function of the AL. The specific form of the resulting curve is determined by the specific values of a_0, a_1, and a_2 (Figure 5-6).

Generally, for a given lens, the numerical values of the three constants (a_0, a_1, a_2) are derived from a double regression analysis of *d* vs AC and AL, where *d* is the optical ACD producing the true postoperative refraction. For this purpose, however, postoperative data must be available. Prior to this, another method to determine a_0, a_1, and a_2 is necessary.

It was found[22] that quite a number of lenses could well be described by fixed values for a_1 and a_2, namely a_1=0.4 and a_2=0.1. Therefore, in "standard mode", we set a_1=0.4, a_2=0.1, and derive a_0 from the manufacturers' ACD constant (ACD_{const}) according to:

$$a_0 = ACD_{const} - 0.4\ mean(AC) - 0.1\ mean\ (AL) \qquad (3)$$

with mean(AC) = 3.37 and mean(AL) = 23.39[20]

Acknowledgments

The author wishes to thank the surgeons Z. Duzanec (University Eye Clinic, Wuerzburg, Germany; Chiron 90D and Rayner 755U) and J. Brändle (Private Practice, Füssen, Germany) for providing patient data.

References

1. Hoffer KJ. The Hoffer Q formula: a comparison of theoretic and regression formulas. *J Cataract Refract Surg.* 1993;19:700-712.
2. Holladay JT, Musgrove KH, Prager TC, Lewis JW, Chandler TY, Ruiz RS. A three-part system for refining intraocular lens power calculations. *J Cataract Refract Surg.* 1988; 14:17-24.
3. Retzlaff J, Sanders DR, Kraff MC. Development of the SRK/T intraocular lens implant power calculation formula. *J Cataract Refract Surg.* 1990;16(3):333-340.
4. Retzlaff J. A new intraocular lens calculation formula. *American Intra-Ocular Implant Society Journal.* 1980;6:148-152.
5. Sanders DR, Retzlaff J, Kraff MC. Comparison of the SRK II formula and other second generation formulas. *J Cataract Refract Surg.* 1988;14:136-141.
6. Holladay JT. Standardizing constants for ultrasonic biometry, keratometry, and intraocular lens power calculation. *J Cataract Refract Surg.* 1997;23:1356-1370.
7. Haigis W. IOL calculation according to Haigis. 1997. Available at: http://www.augenklinik.uni-wuerzburg.de/uslab/ioltxt/haie.htm. Accessed September 1, 2003.
8. Haigis W, Lege B, Miller N, Schneider B. Comparison of immersion ultrasound biometry and partial coherence interferometry for IOL calculation according to Haigis. *Graefes Arch Clin Exp Ophthalmol.* 2000;238:765-773.
9. Hill WE. The Haigis formula. Jan 30, 2003. Available at: http://www.doctor-hill.com/haigis.htm. Accessed September 1, 2003.
10. Haigis W, Duzanec Z, Kammann J, Grehn F. Benefits of using three constants in IOL calculation. Poster at: Joint Meeting, American Academy of Ophthalmology, Pan-American Association of Ophthalmology; October 24, 1999; Orlando, FL. Available at: www.scientificposters.com/aao/
11. Hill WE. The Haigis formula for IOL power calculation. *Geriatric Ophthalmology.* 2002;1(1):8.
12. Retzlaff J, Sanders DR, Kraff MC. *Lens Implant Power Calculation: A Manual for Ophthalmologists & Biometrists.* 3rd ed. Thorofare, NJ: SLACK Incorporated; 1990.
13. Holladay JT. International intraocular lens & implant registry 2000. *J Cataract Refract Surg.* 2000;26:118-134.
14. Norrby NE, Korany G. Prediction of intraocular lens power using the lens haptic plane concept. *J Cataract Refract Surg.* 1997;23(2):254-9.
15. Haigis W, Gross A. Modellrechnungen zur Vorhersage von IOL-Konstanten. In: Vörösmahrty D, Duncker G, Hartmann Ch (Hrsg): *10. Kongress der Deutschen Gesellschaft für Intraokularlinsen-Implantation und Refraktive Chirurgie, Budapest 1996.* Berlin: Springer-Verlag; 1997;288-294.
16. Haigis W. Strahldurchrechnung in Gauß'scher Optik zur Beschreibung des Systems Brille-Kontaktlinse-Hornhaut-Augenlinse (IOL), *4. Kongreß d. Deutschen Ges. f. Intraokularlinsen Implant, Essen 1990.* Berlin: Springer-Verlag; 1991;233-246.
17. Lepper RD, Trier HG. Refraction after intraocular lens implantation: results with a computerized system for ultrasonic biometry and for implant lens power calculation. *Doc Ophthal Proc Ser.* 1983;38:243-248.
18. Olsen TH. Prediction of intraocular lens position after cataract extraction. *J Cataract Refract Surg.* 1986;12:376-379.
19. Olsen TH, Corydon L, Gimbel H. Intraocular lens power calculation with an improved anterior chamber depth prediction algorithm. *J Cataract Refract Surg.* 1995;21:313-319.

20. Haigis W, Waller W, Duzanec Z, Voeske W. Postoperative biometry and keratometry after posterior chamber lens implantation. *Eur J Implant Ref Surg*. 1990;2:191-202.
21. Haigis W. Einfluß der Optikform auf die individuelle Anpassung von Linsenkonstanten zur IOL-Berechnung. In: Rochels R, Duncker GIW, Hartmann CH, eds. *9. Kongress d. Deutsch. Ges. f. Intraokularlinsen Implant, Kiel 1995*. Berlin: Springer-Verlag; 1996;183-189.
22. Haigis W, Kammann J, Dornbach G, Schüttrumpf R. Vorhersage der postoperativen Vorderkammertiefe bei Implantation von PMMA- und Silikonlinsen im Kapselsack. In: *7. Kongr. d. Deutsch. Ges. f. Intraokularlinsen Implant, Zürich 1993*. Berlin: Springer-Verlag; 1993;505-510.
23. Hill WE. *Choosing the right formula*. Available at: http://www.doctor-hill.com/formulas.htm. Retrieved January 30, 2003.

6

The Refractive Formulas

The refractive formulas are based on the refractive error of the eye and do not take into consideration the axial length (AL) measurements. These formulas are basically used in three situations:

1. To calculate the power for a phakic intraocular lens (IOL) that is placed in front of the natural lens of the eye to correct a refractive error.
2. To calculate the power of a secondary IOL used to correct an aphakic eye.
3. To calculate the power of a piggyback IOL inserted in front of an existing IOL to correct a residual ametropia.

> The refractive formulas are based on the refractive error of the eye and do not depend on the AL.

Holladay Refractive Formula

For IOL power calculations in phakic eyes, the optical elements of the eye (cornea, crystalline lens) remain unchanged. Calculation of the required IOL power can be accomplished without a measurement of the AL. The only required measurements are:

K:	The corneal power, in diopters (D)
PreRx:	The spectacle preoperative refraction, in diopters
DpostRx:	The desired postoperative refraction, in diopters
V:	The spectacle vertex distance, in mm
ELP:	The estimated lens position (ELP) where ELP = 3.74 + S factor, in millimeters, and where 3.74 represents an estimate of the anatomic anterior chamber depth (ACD). The S factor can be calculated for a specific IOL using the equation:

$$S \text{ factor} = (A\text{-constant} \times 0.5663) - 65.60$$

The formula is written:[1]

$$P = \cfrac{1336}{\cfrac{1336 - ELP}{\cfrac{1000}{\cfrac{1000}{PreR_x} - V} + K}} - \cfrac{1336}{\cfrac{1336 - ELP}{\cfrac{1000}{\cfrac{1000}{DpostR_x} - V} + K}}$$

Clinical Application

A patient requires a phakic IOL. The following parameters are entered into the formula:
- Preoperative refraction = -8.25 Sph
- Desired postoperative refraction = -0.75 Sph
- Corneal power = 44.12 D
- Vertex distance = 13 mm
- S factor = -1.027 (A-constant of 114)

The power of the IOL needed is calculated to be -8.00 D.

Gills Refractive Equation

Gills uses the following equation to calculate the power of a piggyback implant needed to correct a residual hyperopic refractive error:[2]

$$P = (Error \times 1.4) + 1\ D$$

The disadvantage of this equation is that it does not differentiate between implants that have different A-constants and it does not correct for myopic errors.

Shammas Refractive Equations

Using regression studies, two refractive equations were derived to calculate the power of a piggyback implant required to correct a resultant refractive error.[3] These equations yield the best results for errors ranging from -5.00 D to +5.00 D and have an accuracy of ± 0.50 D. The advantage of these equations is their simplicity. It uses only the refractive error to be corrected and the A-constant of the implant to be used. It requires no measurements of the AL or of the corneal power.

To correct a hyperopic (+) error, the equation is:

$$P = \frac{Error\ (+)}{0.03(138.3 - A)} - 0.50$$

To correct a myopic (-) error, the equation is:

$$P = \frac{Error\ (-)}{0.04(138.3 - A)} - 0.50$$

Clinical Application

After cataract surgery, the operated eye ended with a refractive error of +4.00 D. The surgeon plans to insert an acrylic piggyback implant with an A-constant of 118.4. The power of the implant is calculated:

$$P = \frac{+4.00}{0.03(138.3-118.4)} - 0.50 = 6.70 - 0.50 = 6.20 \text{ D}$$

Clinical Application

A patient requires a phakic IOL. The desired correction is -7.50 D and the implant to be used has an A-constant of 114. The power of the implant is calculated:

$$P = \frac{-7.50}{0.04(138.3-114.0)} - 0.50 = -7.72 - 0.50 = -8.22 \text{ D}$$

References

1. Holladay JT. Refractive power calculations for intraocular lenses in the phakic eye. *Am J Ophthalmol*. 1993;116: 63-66.
2. Perrone DM. Modified intraocular lens power formula in polypseudophakia. *J Cataract Refract Surg*. 1996;22:1392-1393.
3. Shammas HJ. Intraocular lens power calculations. In: Azar DT, ed. *Intraocular Lenses in Cataract and Refractive Surgery*. Philadelphia, PA: WB Saunders; 2001.

Hoffer's Iseikonic Formula

Hoffer's iseikonic formula[6] is written:

$$I = \frac{1336}{L - C - 0.05} - \frac{1.336}{\dfrac{1.336}{K + S} - \dfrac{C + 0.05}{1000}}$$

Where I is the iseikonic IOL power in diopters, L is the axial length of opposite eye in mm (minus 0.657 mm if phakic only), C is the anterior chamber depths in millimeters, K is the corneal power in diopters, and S is the spherical equivalent refractive error in diopters of opposite eye.

In most cases, the operated eye needs to be made 2 D more myopic than the other eye to obtain iseikonia. This is true when an iris-supported or an anterior chamber lens is used, but the formula cannot be used for posterior chamber lenses.

Shammas Iseikonic Equations

The IOL power for iseikonia (Pi) is calculated using the same formula for emmetropia with modifications in the values of the axial length (AL) and of the K readings.[7]

In this formula the axial length L is substituted by Li where:

$$Li = Lopp - 1/3\,(5.85 - C)$$

Where Lopp is the axial length of the opposite phakic eye and C is the estimated postoperative anterior chamber depth.

These changes in the AL are based on calculations done with van der Heijde's nomograms. The concept is to keep the posterior focal length equal in both eyes in order to have the same image size. Theoretically, in an average eye the lens can be assimilated to a thin lens at 5.85 mm behind the corneal vertex. When an implant is inserted, it is located between 2.9 mm (PMMA anterior chamber lens) and 5.7 mm (acrylic posterior chamber lens) behind the corneal vertex. This creates a certain elongation of the posterior focal length of the implanted eye if an emmetropic IOL is used. In order to keep the same posterior focal length as the fellow phakic eye, the power of the implanted IOL has to be stronger, as if it has to focus the image in front of the retina, thus making the implanted eye more myopic. The required additional myopia in the pseudophakic eye in order to keep the same image size as in the fellow phakic eye varies with the type of implant and where it sits in the eye (Table 7-1).

The AL is not changed if the fellow eye is already pseudophakic since both implants will be almost at the same distance from the corneal apex.

The second modification concerns the corneal power. K is changed to Ki where:

$$Ki = K + Ropp$$

Where Ropp is the refraction of the opposite eye at the corneal plane.

TABLE 7-1.

Required Additional Myopia in the Operated Eye When Compared to the Fellow Eye in Order to Achieve Iseikonia

Implant	AC Depth	Additional Myopia
Anterior chamber lens	2.9 mm	-2.50 D
Iris supported lens	3.5 mm	-2.00 D
Plano-convex PMMA lens	4.3 mm	-1.25 D
Biconvex PMMA lens	5.3 mm	-0.40 D
Acrylic PC lens	5.7 mm	-0.15 D

For refractions of -3.0 to +3.0, this number is almost the same as the spectacle refraction. In the presence of a higher refraction, the refraction at the cornea (R) can be calculated from the spectacle correction (R spec):

$$R = R \text{ spec} \times \frac{1}{1 - 0.012 \text{ R spec}}$$

The two equations involving the changes in the AL and in the K readings can be incorporated into any IOL power formula to derive the iseikonic IOL power, and the final refraction achieved with such an IOL.

Clinical Application

A 52-year-old patient is diagnosed with a cataract in her left eye. The right eye only shows incipient cataract changes. The preoperative measurements are as follows:

	OD	OS
Vision	20/25	20/200
Refraction	-6.50 Sph	-6.50 Sph
Axial length	26.85 mm	26.90 mm
K readings	45.25 D	45.50 D
IOL for emmetropia	8.03 D	7.50 D

The decision is to operate on the left cataractous eye, and since the right eye is still phakic and does not require surgery in the near future, it becomes imperative to calculate the iseikonic lens power, the final refraction that will be achieved with such an IOL, and the 5% tolerance range.

1. First we calculate $L_i = L_{opp} - (1/3)(5.85 - C) = 26.85 - (1/3)(5.85 - 5.70)$
   $= 26.80$ mm

2. Then we calculate $K_i = K + R_{opp} = 45.50 - \dfrac{6.50}{1 - (0.012)(-6.50)} = 39.50$ D

Clinical Application, continued

The IOL power for iseikonia is calculated with an axial length (Li) of 26.4 mm, and K readings (Ki) of 39.50 D and found to be 16.0 D.

The final refraction achieved with such an IOL is calculated to be -6.50 Sph. The 5% iseikonic range (approximately ± 2.50 D) will then be from -4.00 Sph to -9.00 Sph. In this specific case, the final refraction should not be below -4.00 Sph to avoid any aniseikonia. At the same time, you also would have avoided any significant anisometropia between the two eyes.

The IOL calculations are then repeated for the left eye using the measured AL for that eye (26.90 mm) and the measured K readings (45.50 D). The IOL needed to produce a final refraction of -4.00 Sph is calculated to be 12.5 D.

Iseikonic calculations are rarely performed these days for the following reasons:
- They require additional complicated calculations.
- They maintain the operated eye highly myopic or highly hyperopic.
- The surgeon has to balance between induced anisometropia and aniseikonia.

Fortunately, iseikonic calculations only have to be performed when the fellow eye is highly myopic or highly hyperopic and does not require surgery.[8]

These calculations become obsolete if the refractive error in the fellow eye is corrected with a contact lens, corneal refractive surgery, or early cataract and implant surgery.

References

1. Binkhorst RD. The optical design of intraocular lens implants. *Ophthalmic Surg.* 1975;6:17-31.
2. Binkhorst RD. *Intraocular Lens Power Calculation Manual: A Guide to The Author's TICC-40 Programs.* 3rd ed. New York: R.D. Binkhorst; 1984.
3. van der Heijde GL. A nomogram for calculating the power of the prepupillary lens in the aphakic eye. *Bibl Ophthalmol.* 1975;83:273-275.
4. van der Heijde GL. The optical correction of unilateral aphakic. *Trans Am Acad Ophthalmol Otolaryngol.* 1976;81:80-88.
5. Huber C, Binkhorst C. Iseikonic lens implantation in anisometropia. *American Intra-Ocular Implant Society Journal.* 1979;5:194-198.
6. Hoffer KJ. Intraocular lens calculations: The problem of the short eye. *Ophthalmic Surgery.* 1981;12:269-272.
7. Shammas HJ. Integrated iseikonic equation for intraocular lens power calculations. *J Cataract Refract Surg.* 1994;20:74-77.
8. Troutman RC. Artiphakia and aniseikonia. *Trans Am Ophthalmol Soc.* 1962;60:590-658.

8

Physical Principles of A-Scan Ultrasound

Shane Dunne, PhD

Sound

The term *ultrasound* refers to sound waves beyond the range of human hearing. In order to make this definition more precise, and to explain the properties of ultrasound, we must first define and explain *sound*.

The 2002 CD-ROM edition of *Encyclopædia Britannica* defines sound as "a mechanical disturbance from a state of equilibrium that propagates through an elastic material medium."[1] This definition is perhaps easiest to understand by looking at it backwards.

An "elastic material" is a material which can be compressed, and which springs back to its original shape when the compressing force is removed. "State of equilibrium" is a fancier name for the original shape. Consider knocking on a door, as shown in Figure 8-1.

When the knuckles of the hand strike the door's surface, the molecules of which the door is made are temporarily forced closer together, ie, the material of the door is compressed. Because the material is elastic, the molecules spring back to their original positions, and in so doing, strike adjacent molecules deeper within the door, forcing them closer together. The compression, or "mechanical disturbance" has thus moved or "propagated" deeper into the "medium", which is the material of the door. This process continues until some molecules at the far side of the door are compressed, and in springing back to their original position, cause compression of the molecules in the air adjacent to the door. This disturbance then propagates through the air, perhaps reaching the ear of a person inside. Disturbances that propagate in this way are generally called *waves*; hence we speak of "sound waves."

Figure 8-1. Propagation of a compression wave (knock) through a medium (door).

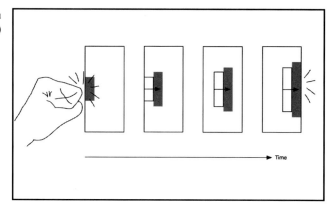

Figure 8-2. Expanding spherical wavefront.

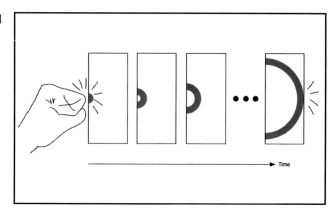

Wavefronts and Sound Sources

Figure 8-1 represents the motion of a compression wave in an oversimplified way, to clarify the point that it is the decompression or "springing back" of one part of the medium that provides the energy to compress an adjacent part, and thus the compression wave moves through the medium, while the individual molecules remain close to their equilibrium positions.

The actual situation is that the compression wave expands symmetrically outward from the point of impact, as shown in Figure 8-2.

The dark semi-circular band in Figure 8-2 represents the expanding hemispherical ridge of increased pressure, which is technically called the wavefront. For any wave, a wavefront is a surface over which the pressure is the constant at a given instant. Sound propagates along lines or "vectors" perpendicular to the wavefront.

A *point source* of sound (eg, the point of impact of the hand on the door in the figures) creates a spherical wavefront, with sound propagating in all directions away from the source. A piston-like sound source creates a quasi-planar wavefront, with sound propagating mostly in a single direction, as shown in Figure 8-3.

A piston-like sound source whose emitting surface is spherical and concave can create a reverse spherical wavefront where the sound energy converges at a focal point, as shown in Figure 8-4.

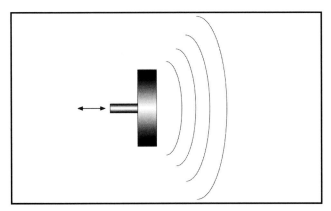

Figure 8-3. Quasi-planar wavefronts created by a broad, flat sound source (piston).

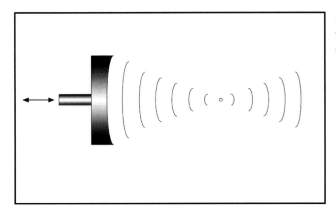

Figure 8-4. Reverse spherical wavefronts created by a spherically focused piston source.

Sound in Time

Look again at Figures 8-3 and 8-4, and suppose that the piston is moving repetitively back and forth at a constant frequency of, let us say, 440 times per second. Plotting a graph of the piston's position against time, we would obtain a curve as shown in Figure 8-5. Such a curve is called *periodic*, because it repeats itself continuously. The smallest repeated portion is called a *cycle*, and the length of time required for one cycle is called the *period*. Most sound phenomena are *aperiodic* (not periodic), but the study of periodic sound provides the theoretical basis for the study of all sound.

A pressure sensing device located anywhere in the path of the sound would detect time variations of pressure in the elastic medium, which, when graphed, would yield a similar curve, where the vertical axis is pressure rather than position. A human ear would sense this periodic sound as having a definite pitch, in this case, A above middle C on the piano. Pitch is the human sensation of sound frequency.

Frequency is measured in cycles per second, also called Hertz (after Heinrich Hertz, a German physicist who studied wave phenomena at the end of the 19th century), which is abbreviated Hz. One kilohertz, abbreviated kHz, equals 1000 cycles per second. One megahertz, abbreviated MHz, equals 1000000 cycles per second. The healthy human ear can detect sound frequencies in a range of about 20 Hz to as much as 20 kHz. Ultrasound is sound at frequencies well above 20 kHz.

Ultrasound is sound at frequencies well above 20kHz.

Figure 8-5. Piston's position plotted against time.

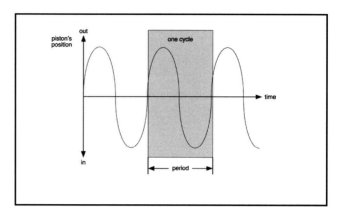

As mentioned earlier, most sound phenomena are not cyclic and are called aperiodic. Aperiodic signals can, however, be analyzed mathematically as though composed of multiple periodic signals of different strengths and frequencies. Real-world ultrasound equipment generates sound pulses whose energy is confined to a limited range or band of frequencies. The size of the range is called the *bandwidth*. The midpoint of the range is called the *center frequency*. Both are typically measured in MHz. A typical A-scan biometry system, for example, might have a center frequency of 10 MHz and a bandwidth of 4 to 6 MHz.

Sound in Space

Imagine for a moment that we had a huge number of very tiny pressure sensors stretched out along a line perpendicular to the wavefronts (ie, along the direction vector of sound propagation) as shown in Figure 8-6.

Each of these sensors would "see" pressure variations that could be graphed over time as in Figure 8-5, but because sound takes a finite time to propagate, the pressure curve seen by any one sensor would be shifted in time relative to that seen by the others. At any *single instant*, however, each sensor would see one specific value of pressure, and if we were to plot these values against sensor position (ie, distance from the source), we would obtain a curve as shown in Figure 8-7.

Note that in Figure 8-7, the horizontal axis represents *distance*, whereas in Figure 8-5, the horizontal axis represents *time*.

The amount of distance corresponding to one cycle is called the *wavelength*, and it depends on both the *frequency* of the sound and the speed or *velocity* at which it propagates through the medium, according to the formula:

$$\lambda = \frac{v}{f}$$

Where λ is the wavelength, v is the velocity, and f is the frequency.

Our 440 Hz sound, moving through air at a velocity of about 330 meters per second, would have a wavelength of 330/440, or about three-quarters of a meter.

Most ocular ultrasound images work at frequencies of 8 to 10 MHz.

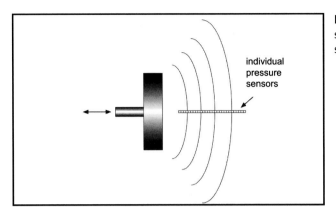

Figure 8-6. Imaginary pressure sensors along the direction of sound propagation.

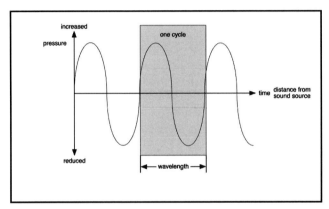

Figure 8-7. Plot of sound pressure versus distance along sound beam, at a single instant.

Most ocular ultrasound images work at frequencies of 8 to 10 MHz. The average velocity of sound in human tissue is about 1550 meters per second. Ten MHz sound in human tissue has a wavelength of 155 microns (millionths of a meter, abbreviated μm).

Interaction of Sound with Media

Consider a sound beam passing from a medium A, into a second medium B, as shown in Figure 8-8. Several phenomena occur at the *interface* (boundary) between the two media. Let v_A denote the velocity of sound in medium A, and let v_B denote the velocity in medium B. For example, medium A might be the aqueous fluid in the anterior chamber, with v_A = 1532 m/sec, and medium B might be the tissue of the crystalline lens, with v_B = 1641 m/sec. As Figure 8-8 shows, part of the sound is *reflected* from the interface, and a part is *transmitted* into medium B. The reflected beam makes the same angle α with the interface normal (a line perpendicular to the interface) as does the incident beam. The transmitted beam makes a different angle β. The specific relationship between α and β isn't all that important*; the key point to understand is that sound does not generally move from one medium to another in a straight line.

---

* The relationship is given by the so-called *law of sines* formula: $\sin \alpha / \sin \beta = v_A / v_B$

Precisely how much of the sound is reflected at an interface, and how much is transmitted, depends on rather complicated acoustic properties of the media*. Again, it's not all that important; the key point to understand is that sound is almost always *partially* reflected and *partially* transmitted at an interface. This simple point is particularly important for ultrasound, because in any A-scan or B-scan ultrasound system, it is reflections that provide the information, but in order to obtain reflections from points deep within the target tissue, at least some of the incident sound must be transmitted through the overlying shallower tissue.

Sound is almost always partially reflected and partially transmitted at an interface.

The interaction of sound and media involves many other specific phenomena, but only two are really important for the proper understanding of A-scan: *scattering* and *absorption*.

Human tissues are not nearly as simple as Figure 8-8 suggests; they are not homogeneous materials with the same speed of sound throughout. The microstructure of tissue (eg, cell structures, local variations of density in the vitreous, etc.) means that any sound beam encounters countless tiny structures, most of which are smaller than the wavelength of the sound itself. For such structures, it is pointless to attempt a reflection/refraction analysis as shown in Figure 8-8; instead we simply say that the sound is *scattered* in multiple directions, and characterize different media according to how strongly they scatter sound.

In addition to scattering sound, real-world media also *absorb* it. When sound energy is absorbed by tissue, it is basically transformed into heat. Absorption depends on many factors, but frequency is one of the more important ones: most media absorb higher frequencies more readily than lower frequencies. This is the reason why most A-scan and B-scan systems, in which the sound must effectively penetrate all the way to the back of the globe and beyond, operate at only 10 or 12 MHz rather than, say, 50 MHz.

For ultrasound systems, the net effect of scattering and absorption is that a large fraction of the sound which goes into the target tissue does not come back out in the form of usable reflections.

A-Scan Echography in Principle

In A-scan echography, an electro-acoustic device called a *transducer* is used as both a source and detector of sound. The transducer is typically mounted at the tip of a handheld probe. In an ideal world, the sound produced by the transducer would be an *impulse*—an infinitely brief burst of increased pressure, rather like the earlier door-knock example. Each time this sound impulse crosses an interface, a similar "echo" impulse would be reflected back and detected by the transducer. To make an A-scan, it would suffice to plot a graph of the detected echo impulses, as shown in Figure 8-9.

In the echograph, each individual echo impulse appears as a "spike" in the graph trace, and so in talking about such graphs (A-scans) we speak of "echospikes". The

---

*The ratio is determined by the difference in *acoustic impedance* of the two media. Acoustic impedance, which is a measure of the relative ease with which sound passes through a medium, depends on many factors beyond the scope of this chapter.

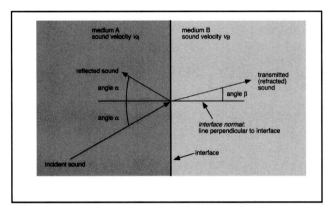

Figure 8-8. Sound interactions at an interface between two dissimilar media.

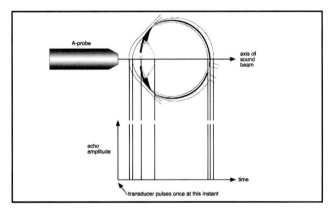

Figure 8-9. Simplified representation of A-scan.

time axis of the graph indicates the "time-of-flight" of the impulse—the total time it takes for the impulse to travel from the transducer, to a given interface, and back to the transducer.

The times at which echo impulses are received can be used to compute the distances between the corresponding interfaces, provided we know the sound velocity. The formula is:

$$d = \frac{tv}{2}$$

Where D is distance, t is the echospike time (taken from the horizontal axis of the echograph in Figure 8-9), and v is the sound velocity.

The factor of 2 occurs because the echospike time t is a time-of-flight measurement of the time required for the sound to travel the distance d twice (outward from the transducer, then back).

A little careful analysis reveals that this formula can be slightly modified to compute distances *between* adjacent interfaces, based on the time *difference* between the corresponding echospikes, using the specific velocity for the intervening medium. For example, the first two echospikes in the graph shown in Figure 8-9 correspond, as shown, to the anterior and posterior surfaces of the cornea. The velocity of sound in corneal tissue has been measured experimentally to be 1641 m/sec. So if the anterior and posterior corneal echospikes occur at points t_a and t_p , respectively, on the echograph time axis, the corneal thickness T_C can be computed as:

$$T_C = \frac{(t_p - t_a) \times 1641}{2}$$

Similarly, the anterior chamber depth (ACD) can be computed from the time between the posterior cornea and anterior lens echospikes using the velocity 1532 m/sec for aqueous; the lens depth can be computed from the time between the anterior and posterior lens spikes using the velocity 1641 m/sec for the natural lens, and the vitreous cavity depth can be computed from the time between the posterior lens and retina spikes using the velocity 1532 m/sec for vitreous. Moreover, we can correct for other media by using the proper velocities, eg, 980 m/sec for silicone IOLs, 2718 m/sec for PMMA IOLs, and so on. Modern A-scan biometers perform such calculations automatically.

A-Scan Echography in Practice

Unfortunately, it is not quite as simple to build an A-scan machine as the previous section implied. For a variety of reasons, no real-world transducer can produce an ideal impulse. What we get instead is a sound pulse of finite duration, whose sound-pressure graph is similar to that shown in Figure 8-10.

Such a pulse can be thought of as a short burst of a single frequency, called the *center frequency* of the transducer, which for ocular ultrasound is typically 8 or 10 MHz. The total burst duration is difficult to define, because the intensity builds up and trails off slowly, but it is customary to consider the interval during which the intensity is at least one-half of the maximum.

Echoes of this pulse will look substantially similar to the pulse itself, ie, they will also have finite duration and a lot of sound-pressure wiggles. To turn these into a nice echograph like that of Figure 8-9, an electronic circuit called an envelope detector is used. Given the pulse shown in Figure 8-10 as input, this circuit will output a voltage signal corresponding to the instantaneous intensity of the echo, as shown in Figure 8-11.

The width or "thickness" of the detector output pulse determines how well the A-scan system can distinguish closely-spaced interfaces—its axial resolution. Many factors combine to determine this pulse width, one of the most important being the bandwidth of the system electronics. An A-scan system with a large bandwidth will produce narrower echospikes—hence higher resolution—than one with a smaller bandwidth.

Figure 8-12 is a revised version of Figure 8-9, with the echograph modified to show the result of using an envelope detector with a rather narrow-bandwidth pulse.

In the echograph of Figure 8-12, note how it is only just possible to distinguish between the anterior and posterior corneal echospikes, and not really possible to distinguish the retinal echospike from the anterior scleral spike. This is an example of the limited axial resolution typical of a real-world A-scan system.

Two other considerations arise in practice: *attenuation* and *noise*. The effects of scattering and absorption in tissue mean that echospikes from distant interfaces tend to be shorter (due to the echosignals themselves being fainter) than those from interfaces closer to the A-probe. We say that the deeper echoes are attenuated with

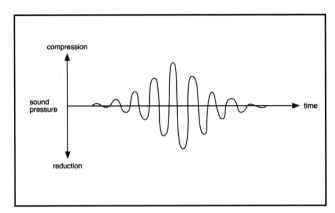

Figure 8-10. Sound-pressure graph of a realistic A-scan pulse.

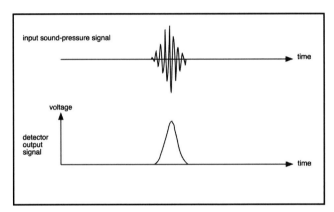

Figure 8-11. Envelope-detector response to a realistic A-scan pulse.

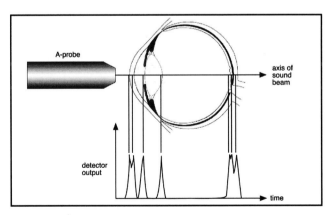

Figure 8-12. More realistic representation of A-scan.

respect to shallower ones. This is not very serious unless the attenuated signals become as weak as the random "background noise" which plagues all sensitive electronics. Figure 8-13 shows the echograph from Figure 8-12 as it might appear when the effects of attenuation and noise are very strong.

Figure 8-13 also shows the large initial spike typically seen at the beginning of an A-scan trace, which is caused by the fact that the echo-receiving circuitry also "sees" the high-voltage pulse used to create the outgoing sound burst.

Figure 8-13. A-scan echograph showing effects of severe attenuation and noise.

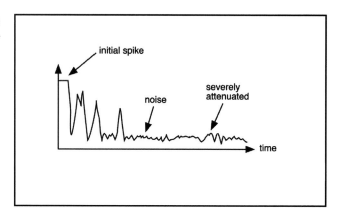

Dynamic Range, Sensitivity, and "Gain"

An A-scan system responds to echoes within a certain intensity range. The faintest detectable echo will be scarcely distinguishable from the background noise level; an example is the retina and scleral echoes in Figure 8-13. At the other extreme, a very strong input signal will overwhelm the electronics, causing the output level to "max out" or clip to the circuit's maximum output level; this is what is happening with the initial spike in Figure 8-13, giving rise to its flat-topped appearance. The difference between these two extreme input levels is known as the *usable dynamic range* of the A-scan system.

Dynamic range, or indeed any comparison of signal levels, is usually specified using the decibel scale. Given any two signal voltages v_1 and v_2, the difference between them in decibels (denoted dB) is given by:

$$20\log \ (v_1/v_2)$$

If v_1 is a stronger signal than v_2, this quantity will be positive. If v_2 is the stronger of the two signals, it will be negative. If the two signals are equal, the decibel difference is zero. When one signal is half as strong as another, the difference is 6 dB. The decibel scale is highly non-linear—a difference of 60 dB implies that one signal is not 10 or 20 times weaker than the other, but rather 1000 times.

An A-scan system requires a usable dynamic range of at least 40 to 50 dB, which means that the echo level above which the receiving circuitry will clip must be at least 40 to 50 dB above the background noise level. A good modern A-scan unit will have a usable dynamic range of at least 60 dB, and a really good one may exceed 80 dB.

Dynamic range is a relative measure; it specifies how much greater the maximum input level is, relative to the minimum distinguishable level, but it does not specify the absolute value of either level. Setting the absolute minimum level defines the sensitivity of the A-scan unit. Technically, sensitivity is quite a difficult quantity to define, but in practice, an A-scan system should ideally be sensitive enough to detect the presence of weak echoes from vitreous structures such as PVD, floaters, and hemorrhage.

Unfortunately, the ultrasound industry has not really settled on a standard definition of sensitivity. This prompted Dr. Karl Ossoinig to establish the discipline of *standardized echography*[2] at the end of the 1970s. The results of this work have still not yet propagated to A-scan biometry.

The earliest analog electronic biometers featured an adjustable-gain amplifier as part of the echo receiving circuitry; increasing the gain had the effect of increasing sensitivity*. Amplifier gain, which is defined as the difference in strength of the output signal relative to the input, is also measured in decibels, and hence the gain setting of many biometers tended to be calibrated in dB. It is important to note, however, that a gain setting of, say, "50 dB" on one instrument did not usually correspond (in terms of equivalent sensitivity, dynamic range, or any other measure) to a gain setting of "50 dB" on any other instrument of a different make or model.

Linear, Logarithmic, and S-Curve Response

The earliest A-scan instruments produced echographs using a linear intensity scale (the vertical axis in Figures 8-9 through 8-13). The result is a very clean-looking graph trace with thin, well-defined echospikes, such as that shown in Figure 8-14. Unfortunately, it is difficult to achieve a large, usable dynamic range with linear scaling.

A much larger usable dynamic range can be achieved by using logarithmic amplifier circuitry, which supports an A-scan display with a logarithmic intensity scale. As Figure 8-15 illustrates, the resulting echographs feature spikes which appear wider, but allow better definition of weak echoes, eg, incipient cataracts and vitreous irregularities.

The work on standardized echography gave rise to the so-called S-curve scale, which combines the clarity benefits of linear scaling with the increased dynamic range of logarithmic scaling. Essentially, the S-curve scale expands the mid-range portion of the display and compresses the upper and lower portions. Several newer biometry instruments make use of S-curve scaling. Figure 8-16 shows an example.

---

* Only up to a point. Raising the gain too far would cause clipping of echospikes and overwhelm the display with the resulting amplified electronic noise.

9

The A-Scan Biometer

A-scan biometers are instruments used to measure the axial length (AL). Most units are compact, efficient, computerized, and complete with IOL power calculation capabilities, and a print-out of the measurements and the results. Biometry is also provided on larger ultrasound units used for diagnostic A- and B-scan examinations.

Biometry units vary from simple to highly sophisticated. Eight different components are discussed herein: the probe configuration, the transducer, the emitted sound beam, the oscilloscope, the sensitivity setting, the velocity setting, the electronic gates, and the biometer's report. The clinical importance of each component is also discussed.

The Probe

The probe is connected to the main chassis of the biometer by an electronically shielded cable and contains a transducer at its tip.[1-3] The original solid probe (Figure 9-1), such as the one available with the Kretz 7200 MA ultrasound unit (GE Medical Systems, Waukesha, WI), is basically used to measure the AL through an immersion technique. This probe is also used for standardized A-scan echography.

The newer and thinner solid probes (Figure 9-2) are becoming more popular for AL measurement. They can be used for immersion technique in the same manner as the Kretz solid probes. Furthermore, when used for a contact technique, they produce minimal corneal indentation. Some probes (Figure 9-3) are designed to attach to the tonometer holder of the slit-lamp for better control.

The semi-soft probe has a membrane at the tapered end of a cylinder fastened with an o-ring. These semi-soft probes were once very popular, but they require a certain amount of maintenance, such as a regular exchange of the membrane, the o-ring, and the water within the cylinder. Many ultrasound companies have phased

The thin solid probes can be used for both immersion and contact biometry.

Figure 9-1. Large solid probe available on the Kretz 7200 MA ultrasound unit. Nowadays, this probe is rarely being used for biometry.

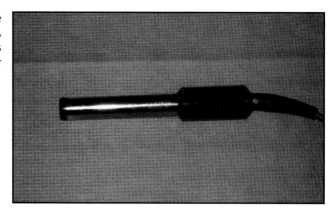

Figure 9-2. Thinner solid probe available on the new I³ biometry unit (Innovative Imaging Inc., Sacramento, CA). At present, this size probe is the most commonly used for biometry.

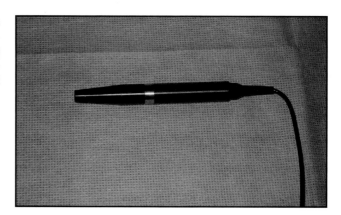

Figure 9-3. Small solid probe that attaches to the tonometer holder.

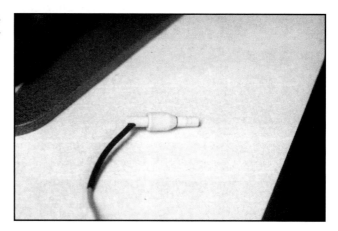

out these semi-soft probes and replaced them with the smaller, less cumbersome compact solid probes.

The Transducer

The transducer consists of a quartz crystal located at the tip of the probe.[1-3] The crystal has piezo-electric qualities that transform electric energy into mechanical energy in the form of sound waves, and vice-versa (Figure 9-4).

The crystal is connected by two electrodes to the main ultrasound system. Damping material is present behind the crystal to minimize reverberations of the ultrasound energy within the probe. Based on the piezo-electric principle, the transducer is an ultrasound source emitting a pulsed ultrasound beam. The performance of the crystal depends mainly on its shape, its diameter, and its thickness.

Clinical Application

1. A transducer with a flat anterior surface will emit a non-focused beam (Figure 9-5). This non-focused beam is an integral part of standardized diagnostic A-scan echography. During examination, the system sensitivity is decreased by 10 to 20 decibels (dB); the beam width decreases for a more accurate AL measurement. The non-focused beam with its parallel borders allows accurate pattern recognition in the short, normal, and long eyes due to its consistently narrow beam.

2. A transducer with a concave surface emits a focused ultrasound beam (Figure 9-6) essential for diagnostic B-scan echography because the retinal layer is located within the beam's focal zone. Ultrasound manufacturers, using the popular B-scan technology, adopted the focused ultrasound beam within their biometry units. When a normal eye is measured, the retina is in the focal zone of the ultrasound beam, yielding an accurate measurement. When a very short or a very long eye is measured, the retina is no longer in the focal zone of the ultrasound beam, causing difficulties in the pattern recognition and a less accurate measurement.

The Ultrasound Beam

Ultrasound consists of high frequency sound waves over 20000 cycles per second, which is the highest frequency audible to the human ear.[4] The ultrasound beam is formed of ultrasound waves that display different characteristics depending on the ultrasound frequency, wavelength, velocity, and direction.

The *frequency*[5] is the number of hertz (Hz) or cycles per second. Higher frequencies provide a higher resolution, while lower frequencies provide better penetration, but a reduction in the resolution. To obtain the high resolution needed for AL measurement, biometry units use ultrasound frequencies ranging between 8 and 25 MHz (1 MHz = 1 megahertz = 1 million cycles per second).

Figure 9-4. The piezo-electric principle is based on the fact that changes in the polarity of an electric current passing through a quartz crystal will cause changes in the shape and size of this crystal, and vice versa. When an electric current passes through the quartz crystal, its thickness decreases or increases according to the polarity of the current. This will transform the electric energy into mechanical energy in the form of sound waves. On the other hand, when the sound waves return to the probe, the mechanical energy will modify the thickness of the crystal, which in turn affects the electric charges at its surface, producing electric energy.

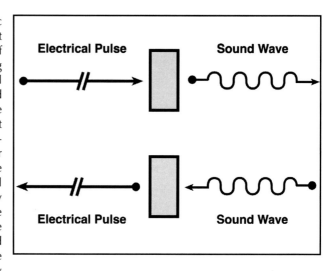

Figure 9-5. Non-focused beam emitted by a transducer (T) with a flat anterior surface.

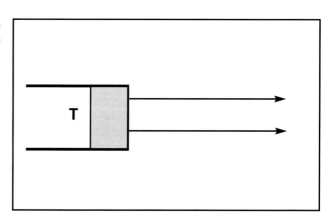

Figure 9-6. Focused beam emitted by a transducer (T) with a concave anterior surface.

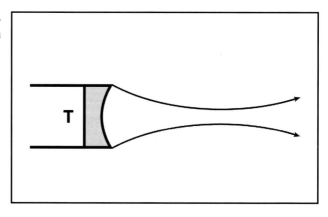

The *wavelength*[6] is the distance between two particles in the same phase of oscillation. Within the ocular tissues, the wavelength is approximately 0.19 mm if an 8 MHz probe is used and 0.15 mm if a 10 MHz probe is used.

The *velocity*[7] is the speed of sound propagation and is expressed in meters per second (m/s). The velocity varies according to the medium through which sound propagates; within the eye, the ultrasound velocity is 1532 m/s in aqueous and vitreous, 1640 m/s in a clear lens, and 1550 m/s in solid tissues. During an accurate measurement of the different eye components, the proper sound speed has to be used for each of these entities.

The *direction* of the ultrasound beam[8,9] affects the display of the tissues under examination. During biometry, the emitted sound beam will meet multiple interfaces. At each interface, part of the sound beam is reflected back toward the probe, and the remainder of the sound beam keeps propagating deeper into the tissues. This process will generate echospikes from the different interfaces that have been intersected, ie, anterior surface of the cornea, posterior surface of the cornea, anterior surface of the lens, posterior surface of the lens, anterior surface of the retina, and anterior surface of the sclera. When the ultrasound beam reaches the orbital tissues, it is attenuated until it loses all its energy. The sound beam returns to the transducer that also acts as a receiver. The pulses are then processed within the biometer to display "echosignals" on the screen.

Clinical Application

During AL measurement, it is important to display "maximal echospikes" by keeping the incident beam perpendicular to the different interfaces. Only then will the returning beam carry the maximal energy reflected by the tissues. This is especially true about the retinal spike.

When the probe is aligned with the optical axis of the eye, the sound beam is perpendicular to the retinal interface and the portion of the returning beam that reaches the probe will carry maximal energy from the surface under examination (ie, the retina) and the retinal spike is displayed as a straight, steeply-rising echosignal (Figure 9-7).

When the probe is not properly aligned with the optical axis of the eye, the sound beam becomes tangential to the retinal surface (instead of being perpendicular to it) and the portion of the returning beam that reaches the probe will carry minimal energy from the surface under examination (ie, the retina). The retinal spike is displayed as a jagged, slow-rising echospike (Figure 9-8).

The Oscilloscope

The oscilloscope is the display screen that allows the technician to monitor the different echospikes generated by the ocular surfaces. Most units are equipped with an automatic image freezing capability when a valid measurement is detected and with gates for an electronic measurement of the AL from the echogram displayed on the screen.

Clinical Application

Units without oscilloscope are not recommended since they do not allow pattern recognition.

Figure 9-7. When the ultrasound beam is perpendicular to the retina, the retinal spike (arrow) is displayed as a straight, steeply-rising echo-spike.

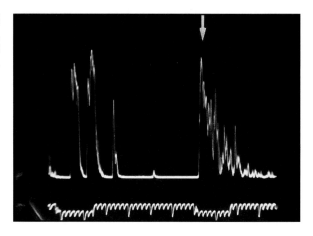

Figure 9-8. When the ultrasound beam is not perpendicular to the retinal surface, the retinal spike is displayed as a jagged, slow-rising echospike.

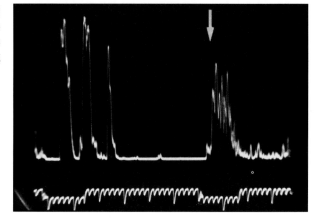

The Sensitivity Setting

The sensitivity setting controls the height of the echospikes displayed on the oscilloscope.

Clinical Application

Each ultrasound unit needs to have its system sensitivity adjusted for better pattern recognition of the anterior and posterior corneal surfaces, anterior and posterior lens surfaces, and anterior retinal surface (Figure 9-9).

The Velocity Setting

The velocity setting controls the speed of sound propagation. The velocity, measured in m/s, varies according to the medium through which sound propagates.

Clinical Application

Most units use an average velocity of 1548 m/s to 1556 m/s in a cataractous eye and 1532 m/s in an aphakic eye. Certain units measure each ocular compartment at its correct velocity; the anterior chamber depth (ACD) is measured with a velocity of 1532 m/s; the cataractous lens is measured with an average velocity of 1640 m/s; depth of the vitreous cavity is measured with a velocity of 1532 m/s like the aqueous. These measurements are then computed within the instrument to display one AL reading.

The Electronic Gates

Electronic gates allow ultrasound units to provide an electronic read-out of the AL in millimeters. Biometers are equipped with two or four gates. The two main gates are the "corneal gate" placed in the region of the corneal spike and the "retinal gate" placed in the region of the retinal spike (Figure 9-10). Such instruments measure the travel time between the anterior surface of the cornea and the anterior surface of the retina and use an average sound velocity for the measurement of the AL.

Instruments equipped with four gates allow the positioning of these gates over the leading edges (Figure 9-11) of the echoes generated from the anterior surface of the cornea, the anterior surface of the lens, posterior surface of the cornea, the anterior surface of the lens, the posterior surface of the lens, and the anterior surface of the retina. A measurement of the ACD, lens thickness, and the total AL are displayed on the screen.

The Print-Out

Most biometers are equipped with built-in minicomputers that analyze the A-scan echograms and also perform IOL power calculations using one or more formulas programmed within the biometer. Once the calculations are done, a print-out is obtained for permanent record.

When a measurement is taken, the A-scan echogram is displayed on the screen. This echogram is printed and kept on file. The advantages of such a practice are:

1. It allows the technician and/or the surgeon to review the echograms prior to surgery. If the measurement is not acceptable, it can be repeated.
2. It keeps a permanent record of the echogram to be reviewed at a later date if the final result is not satisfactory.

Most formulas are programmed within the biometer or are available on calculators and computer diskettes to yield some or all of the following information:

- The IOL power for emmetropia.
- The IOL power to produce a refraction of -1.0, -2.0, or -3.0 D.
- The IOL power to produce any amount of desired ametropia.
- The expected postoperative refractive error with any IOL.
- The IOL power with different formulas.

4. The presence of a diagnostic B-scan and/or pachymeter within the same chassis. Practices that handle vitreoretinal pathology and/or perform corneal surgery will enjoy the added features in one compact unit.

References

1. Buschman W. Special transducer probes for diagnostic ultrasonography of the eyeball. In: Goldberg R, Sarin L, eds. *Ultrasonics in Ophthalmology*. Philadelphia, PA: WB Saunders Co.; 1967.
2. Coleman DJ, Carlin B. Transducer alignment and electronic measurement of visual axis dimensions in the human eye using time amplitude ultrasound. In: Oksala A, Gernet H, eds. *Ultrasonics in Ophthalmology*. Basel, Switzerland: S. Karger; 1967.
3. Gordon D. Transducer design for ultrasonic ophthalmology. In: Gitter K, Keeney A, Sarin L, Meyer D, eds. *Ophthalmic Ultrasound*. St. Louis, MO: CV Mosby Co.; 1969.
4. Shammas HJ. *Atlas of Ophthalmic Ultrasonography and Biometry*. St. Louis: C.V. Mosby Co; 1984.
5. Herrman G, Buschmann W. Methods of measuring the HF oscillation frequency in ultrasound pulses of equipment for diagnostic ultrasonography. *Ophthalmol Research*. 1972;3:274-282.
6. Till P. Solid tissue model for the standardization of the echo-ophthalmograph 7200 MA (Kretztechnik). *Doc Ophthalmol*. 1976;41:205-240.
7. Ludwig GD. The velocity of sound through tissues and the acoustic impedence of tissue. *J Acoust Soc Am*. 1950;22:862-866.
8. Lizzi F, Burt W, Coleman DJ. Effects of ocular structures on the propagation of ultrasound in the eye. *Arch Ophthalmol*. 1970;84:635-640.
9. Lowe R. Time amplitude ultrasonography for ocular biometry. *Am J Ophthalmol*. 1968;66:913-918.
10. Holladay JT, Prager TC, Chandler TY, Musgrove KH. A three-part system for refining intraocular lens power calculations. *J Cataract Refract Surg*. 1988;14:17-24.

A-Scan Biometry

The axial length (AL) is conventionally measured with ultrasonography, using a biometry unit. Measurement of the AL is achieved using an immersion or a contact technique. It is always advisable to have both eyes measured for comparison purposes.

Immersion A-Scan Biometry

TECHNIQUE

The technique described herein can be used with any ultrasound unit equipped with a solid A-scan probe and mobile electronic gates.[1-4]

1. The patient is placed in a supine position on a flat examination table or in a reclining examination chair and a drop of local anesthetic is instilled in both eyes.

2. A scleral shell is applied to the eye. The most commonly used scleral shells are the Hansen shells (Hansen Ophthalmic Development Laboratory, Coralville, IA) (Figure 10-1A) and the Prager shells (ESI, Inc., Plymouth, MN) (Figure 10-1B). The Hansen shells are available in 16, 18, 20, 22 and 24 mm diameter. While the 20 mm shell fits most eyes, the larger cup provides a better fit in bigger eyes with large palpebral fissures and the smaller cups fit better in the presence of a narrow palpebral fissure. The newest Prager shell features single handed immersion biometry, a Luer fitting to facilitate tubing changes, an autostop for exact manufacturer specified probe depth, and six centering guides to ensure perpendicularity. Each shell is polished, allowing direct visualization of fluid levels. Other types of scleral shells are also available from different manufacturers including the Kohn shell (Innovative Imaging Inc., Sacramento, CA) (Figure 10-2). The Prager shell and the Kohn shell have made immersion biometry easier. The use of the Kohn shell is discussed in Chapter 11. In this section, we will describe the routine method used with the Hansen shells.

Figure 10-1A. The Hansen scleral shells are available in 16, 18, 20, 22, and 24 mm.

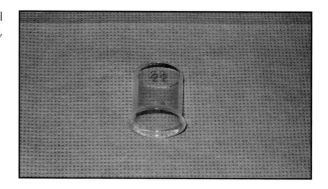

Figure 10-1B. The Prager Shell. In peer reviewed articles, this shell has been shown to eliminate corneal compression associated with the applanation contact technique, and clinically is as accurate as the non-contact inter-ferometer method (IOLMaster). Photo courtesy of Thomas Prager, PhD.

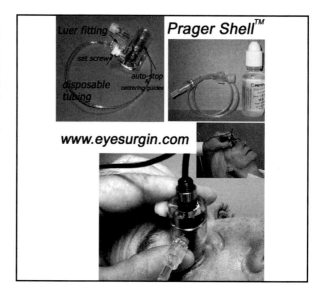

3. The Hansen shell is filled with gonioscopic solution. Methylcellulose 1% is pre-ferred over the 2.5% concentration (too thick) and over saline solutions (too liquid). The solution should be free of air bubbles; the presence of bubbles causes variations in the speed of sound and is responsible for noise formation within the ultrasound pattern. The easiest way to avoid bubbles is to remove the bottle's nipple and to pour the solution in the cup (Figure 10-3). If bubbles do form within the solution, they are removed with a syringe, and, if unsuc-cessful, the cup has to be emptied, cleaned, repositioned, and refilled with gonioscopic solution. The Kohn shell is designed to hold the probe tightly and allow a better fit on the eye. Because of this tight fit, the coupling fluid used in this shell does not have to be methylcellulose; instead, balanced salt solution or artificial tears could be used.

4. The ultrasound probe is immersed in the solution, keeping it 5 to 10 mm away from the cornea (Figure 10-4). The patient is asked to look, with the fellow eye, at a fixation point placed at the ceiling. Attention is then focused on the screen. The probe is gently moved until it is properly aligned with the optical axis of the eye and an acceptable A-scan echogram is displayed on the screen.

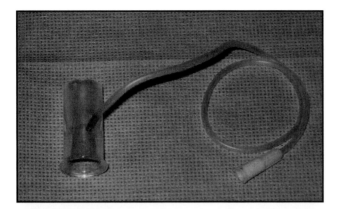

Figure 10-2. The Kohn shell.

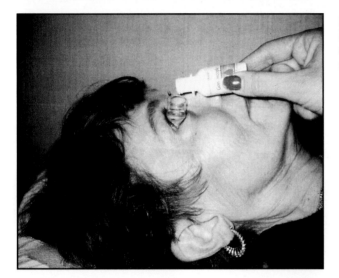

Figure 10-3. The cup is placed between the lids and methylcellulose 1% is poured into the cup.

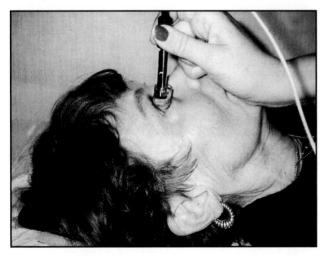

Figure 10-4. The ultrasound probe is immersed in the solution, keeping it 5 to 10 mm away from the cornea.

A-SCAN PATTERN OF THE PHAKIC EYE

The A-scan pattern of a normal phakic eye examined with an immersion technique displays the following echospikes from left to right (Figure 10-5):

IS: The initial spike (IS) is produced at the tip of the probe. It has no clinical significance. Many units will allow the technician to move the whole A-scan pattern to the left and remove the IS from the picture (Figure 10-6).

C: The corneal spike (C) is double peaked, representing the anterior (C1) and posterior (C2) surfaces of the cornea.

L1: The anterior lens spike (L1) is generated from the anterior surface of the lens.

L2: The posterior lens spike (L2) is generated from the posterior surface of the lens, and is usually smaller than L1.

R: The retinal spike (R) is generated from the anterior surface of the retina. It is straight, high-reflective, and tall whenever the ultrasound beam is perpendicular to the retina, as it should be during AL measurement.

S: The scleral spike (S) is another high-reflective spike generated from the scleral surface, right behind the retinal spike, and should not be confused with it.

O: The orbital spikes (O) are low-reflective behind the scleral spike.

The eye is divided ultrasonically into three compartments (See Figures 10-5 and 10-6):

1. The anterior chamber depth (ACD)[5,6] is measured between the anterior corneal surface (C1) and the anterior lens surface (L1) using a velocity of 1532 m/s. If needed, the corneal thickness[5] is measured between the anterior (C1) and posterior (C2) surfaces of the cornea using a velocity of 1620 m/s. The thickness of a normal cornea is approximately 0.5 mm.

2. The lens thickness[7-12] is measured between the anterior lens surface (L1) and the posterior lens surface (L2) using a velocity of 1641 m/s. Actually 1640.5 m/s is the calculated sound velocity in the normal crystalline lens. The sound velocity varies in cataractous eyes with a slower velocity (average 1590 m/s) in the intumescent cataracts due to their high water content, and a higher velocity in the posterior capsular cataracts. In most cases of nuclear sclerosis with or without subcapsular changes, the sound velocity averages 1641 m/s.

3. The vitreous cavity's depth[6] is measured between the posterior lens surface (L2) and the anterior surface of the retina (R) using a velocity of 1532 m/s.

A manual measurement of the axial length[13] is used with older ultrasound units not equipped with an electronic read-out, and is rarely used nowadays. Using calipers, a measurement is taken from the scale in microseconds. However, this reading represents the travel time it takes the ultrasound beam to reach the tissue under examination and return to the probe, thus representing twice the actual measurement. It is then divided by 2, and converted to millimeters using an average velocity.

Some biometers give the readings directly in millimeters using an average sound velocity. This velocity is reported in meters per second (m/s). Most biometers use an average velocity of 1550 to 1555 m/s. A velocity of 1553 m/s is recommended.[14] Most modern biometers use separate sound velocities for the different eye components. The biometer provides an ACD, the lens thickness, and the total AL.[13]

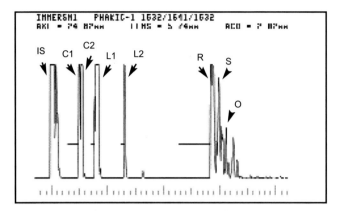

Figure 10-5. Ultrasound display of the different echospikes during immersion biometry, identifying the initial spike (IS), the anterior (C1) and posterior (C2) corneal surfaces, the anterior (L1) and posterior (L2) lens surfaces, the retina (R), sclera (S), and orbital tissues (O).

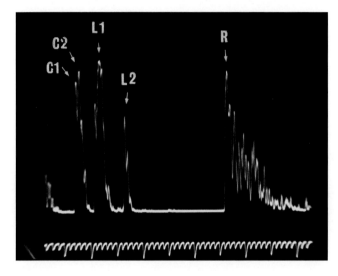

Figure 10-6. A-scan display of a phakic eye where the initial spike has been removed from the screen display and identifying the corneal (C1 & C2), lens (L1 & L2) and retinal (R) spikes.

Contact A-Scan Biometry

TECHNIQUE

The contact technique for AL measurement is an alternative to immersion A-scan biometry. It does not use a scleral shell. Instead, the probe comes in contact with the cornea to generate the first echospike.

1. A drop of local anesthetic is instilled in both eyes. The patient is examined in the seated position, with the chin correctly positioned on a free-standing chin rest.[15-17]

2. The ultrasound probe is attached to a zero-weight balance glide to prevent any pressure on the eye during examination. The probe's movements are controlled through a joystick handle (Figure 10-7). The probe is positioned in front of the eye and the patient is asked to fixate the red light within the probe. The probe is then brought forward to gently touch the cornea without indenting it.

Figure 10-7. Contact A-scan biometry. The technician uses the joystick to align the probe with the patient's eye.

Figure 10-8. The small solid probe fits in the tonometer holder for A-scan biometry.

3. Attention is then focused on the screen. The probe is moved slightly, up and down or to the side to optimize the echospikes displayed on the oscilloscope. A Polaroid picture or a print-out is obtained.

Other probes are mounted on a Goldmann tonometer holder and are designed to be used with a slit-lamp (Figure 10-8). The examination is performed in the same manner as previously described. Hand-held compact biometric rulers such as the Bio-Pen[18] (Mentor O & O, Norwell, MA) have lost some of their popularity and are now rarely used.

Probes used with the contact technique should be cleaned after each examination to prevent contamination.

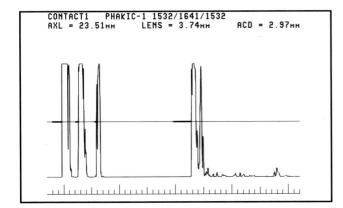

Figure 10-9. A-scan display of a phakic eye measured with contact A-scan biometry. Since the probe is in contact with the eye, the initial spike and the anterior corneal spike become one.

A-Scan Pattern

The A-scan pattern of a phakic eye examined with the contact technique (Figure 10-9) demonstrates similar echospikes, except that the corneal spike is merged with the initial spike since the cornea is in touch with the tip of the probe.

Variations in Axial Length Measurement

Variations in AL measurement are due to the use of different examination techniques and/or to the use of different sound velocities by the biometer.

Variations Due to the Examination Technique

The contact method for AL measurement does not yield the same results as the high precision A-scan biometry. When measuring the same eye, the contact technique yields a shorter measurement than the immersion technique.[18-21] In a prospective study on 180 eyes performed by the author,[21] AL measurements were obtained on each eye with both contact and immersion techniques. Each eye was measured with the Ocuscan-DBR (contact) (Alcon, Irvine, CA), the Ocuscan-400 (immersion) (Alcon), and the Kretz 7200 MA (immersion) units. AL measurements obtained with the contact technique were shorter than measurements obtained with the immersion technique by an average of 0.24 mm.

The two methods of examination differ in the patient's position and the possible corneal applanation by the ultrasound probe. The patient is conventionally examined in the seated position with the contact technique and the probe is brought forward to touch the cornea. The patient is conventionally examined in the supine position with the immersion technique and the solid probe is kept 5 mm to 10 mm away from the cornea. These differences in the methods of examination, mainly the corneal indentation and the subsequent shallowing of the anterior chamber, are responsible for the shorter measurement obtained with the contact technique.

Figure 10-10A. A-scan display of a phakic eye using immersion biometry.

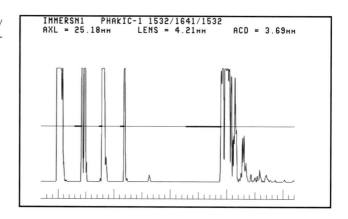

```
IMMERSN1   PHAKIC-1 1532/1641/1532
AXL = 25.18MM        LENS = 4.21MM        ACD = 3.69MM
```

Figure 10-10B. A-scan display of the same phakic eye using contact biometry. Note the shallower anterior chamber depth (ACD) and shorter axial length (AXL) with this technique.

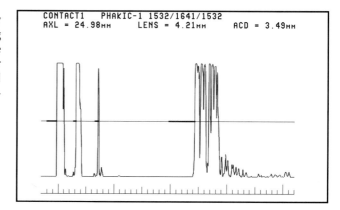

```
CONTACT1   PHAKIC-1 1532/1641/1532
AXL = 24.98MM        LENS = 4.21MM        ACD = 3.49MM
```

Clinical Application

Figure 10-10 shows measurements obtained from the same eye with an immersion technique (Figure 10-10A) and a contact technique (Figure 10-10B). Note the shallower ACD (3.49 instead of 3.69 mm) and shorter AL (24.98 instead of 25.18 mm) obtained with the contact technique.

VARIATIONS DUE TO THE USE OF AN AVERAGE SOUND VELOCITY VERSUS SPECIFIC SOUND VELOCITIES

Although it is best to measure the different ocular compartments at their specific sound velocities, the use of an average sound velocity of 1553 m/s yields clinically insignificant errors in the measurement.

- In the average 23.5 mm eye, the use of an average 1553 m/s sound velocity will yield the exact same measurement as the use of specific sound velocities for the aqueous, lens, and vitreous; this is true if the ACD and lens thickness have average values. The use of slightly lower (1548 m/s) or slightly higher (1556 m/s) sound velocities varies the AL by less than 0.05 mm.

- In the long 26.0 mm eye, with an anterior chamber of 3.2 mm and a lens thickness of 4.2 mm, the use of the average 1553 m/s yields a measurement of 26.04 mm, with an error of only 0.04 mm.
- In the short 21.0 mm eye with an anterior chamber of 2.5 mm and a lens thickness of 4.8 mm, the use of the average 1553 m/s yields 20.93 mm, with an error of only 0.07 mm.

APPROXIMATION OF AN AXIAL LENGTH

Occasionally, the eye to be operated on is difficult to measure, whether it's a recently traumatized eye or an eye with vitreoretinal pathology where the retina cannot be identified. The AL can be approximated by comparing both eyes:
- The opposite eye is measured.
- The measurement is adjusted for the difference in refraction, provided of course that there is no difference in the corneal curvature. The original refraction must be used for accuracy.
- Add 0.1 mm to the AL for each -0.25 D of additional myopia or subtract 0.1 mm for each +0.25 D of additional hyperopia.
- If the eye had a scleral buckle for treatment of a detached retina, add 0.3 mm to the AL.

In most cases where the AL cannot be measured and has to be approximated, the eye is compromised and the chances of visual recovery to the 20/40 level are quite poor.

EFFECTS OF AXIAL LENGTH VARIATIONS ON THE FINAL REFRACTION

Variations in AL measurement affect the final refraction differently in the average, long, and short eyes.[4]
- In an average 23.5 mm eye, a 0.1 mm difference in AL measurement affects the final postoperative refraction by 0.25 D.
- In a long 26.0 mm eye, a 0.1 mm difference in the AL measurement affects the final postoperative refraction by only 0.20 D.
- In the short 21.0 mm eye, a 0.1 mm difference in the AL measurement affects the final postoperative refraction by 0.31 D.

Although the immersion and contact techniques yield consistent and reproducible measurements, the routine measurement of the AL using an immersion technique and the correct sound velocities has been very rewarding, especially in the short and long eyes.

References

1. Byrne SF. Standardized echography, Part I: A-scan examination procedures. *Int Ophthalmol Clin.* 1979;19:267-281.
2. Ossoinig KC. Standardized echography: basic principles, clinical applications and results. *Int Ophthal Clin.* 1979;19:127-285.

3. Shammas HJ. Axial length measurement and its relation to intraocular lens power calculations. *Am Intraocular Implant Soc J*. 1982;8:346-349.
4. Kendall CJ. *Ophthalmic Echography*. Thorofare, NJ: SLACK Incorporated; 1990: 57-106.
5. Delmarcelle Y, Colignon-Brach J, Luyckx-Bacus J. The role of the cornea and lens in biometry of the anterior chamber in normal subjects. *Arch Ophthalmol*. 1970;30:291-300.
6. Bell DS. Approximate determination of the speed of sound through liquids and solids. *J Acoust Soc Am*. 1968;44:656.
7. Oksala A, Lehtinen A. Measurement of the velocity of sound in some parts of the eye. *Acta Ophthalmol*. 1958;36:633-639.
8. Jansson F, Kock E. Determination of the velocity of ultrasound in the human lens and vitreous. *Acta Ophthalmol*. 1962;40:420-433.
9. Coleman DJ, Lizzi FL, Franzen LA, Abramson DH. A determination of the velocity of ultrasound in cataractous lenses. *Bibl Ophthalmol*. 1975;83:246-251.
10. Laursen AB, Fledelius H. Variations of lens thickness in relation to biomicroscopic types of human senile cataract. *Acta Ophthalmol*. 1979;57:1-13.
11. Pallikaris I, Gruber H. Determination of sound velocity in different forms of cataracts. *Documenta Ophthalmologica*. 1981;29:165-169.
12. Massin M, Lambrinakis I. In vivo determination of the speed of ultrasound in cataractous lenses. In: Sampaolesi R, ed.*Ultrasonography in Ophthalmology*. 1990;12:131-134.
13. Shammas HJ. Manual versus electronic measurement of the axial length. In: Hillman JS, LeMay MM, eds. *Ultrasonography in Ophthalmology. Proceedings of the 1982 Ninth SIDUO Congress*. The Hague: Dr. W. Junk Publishers; 1983;225-229.
14. Shammas HJ. A-scan biometry of 1000 cataractous eyes. In: Ossoining KC, ed. *Ophthalmic Echography. Proceedings of the 10th SIDUO Congress*. Dr. W. Junk Publishers. Documenta Ophthalmologica Series. 1987;48:57-63.
15. Binkhorst RD. Biometric A-scan ultrasonography and intraocular lens power calculation. In: Emery JM, ed. *Current Concepts In Cataract Surgery: Selected Proceedings of the Fifth Biennial Cataract Surgical Congress*. St. Louis, MO: CV Mosby; 1987.
16. Coleman DJ. Ophthalmic biometry using ultrasound. *Int Ophthalmol Clin*. 1969;9:667-683.
17. Coleman DJ, Carlin B. A new system for visual axis measurements in the human eye using ultrasound. *Arch Ophthalmol*. 1967;77:124-127.
18. Shammas HJ, Swearingen, M. Clinical evaluation of the Bio-Pen for axial length measurement. *J Cataract Refract Surg*. 1990;16:120-122.
19. Olsen T, Nielsen PJ. Immersion versus contact technique in the measurement of axial length by ultrasound. *Acta Ophthalmologica*. 1989;67:101-102.
20. Schelenz J, Kammann J. Comparison of contact and immersion techniques for axial length measurement and implant power calculation. *J Cataract Refract Surg*. 1989;15:425-428.
21. Shammas HJ. A comparison of immersion and contact techniques for axial length measurement. *American Intra-Ocular Implant Society Journal*. 1984;10:444-447.

High Precision A-Scan Biometry Using the Kohn Immersion Shell

Gus Kohn, CRA, COT, ROUB

The main concern of many physicians is to automate their patient exams and attain some consistency in the resultant diagnostic testing results. They wish to accomplish this by relying on the available technology to provide them with accurate and easily obtained data. The physician should compare the measurements of automated equipment to not only his or her own measurements with a non-automated instrument, but to that of the technician on a non-automated instrument. This ensures consistency and accuracy.

New technology such as the optical coherence biometer (OCB) has been compared to immersion biometry (becoming known as high precision biometry) and determined to be *as good as* the "gold standard" of immersion biometry but not better. Measurements can only be as good as the yardstick you use to calibrate with, never better. Therefore, if immersion biometry and optical coherence biometry are equally accurate, one must look at other factors to decide which method to use. If you choose to use the OCB, you must also have a biometer for those patients with moderate to severe cataracts, because the OCB cannot measure them. On the other hand, immersion biometry is able to measure all patients.

High precision A-scan biometry using the Kohn immersion shell (Innovative Imaging Inc., Sacramento, CA) is a highly reliable method to obtain consistently accurate axial length (AL) measurements on *all* patients.

Figure 11-4. High precision A-scan echogram of a phakic eye showing (from left to right): echoes generated from the probe tip, cornea, anterior lens surface, posterior lens surface, retina, sclera, and orbit.

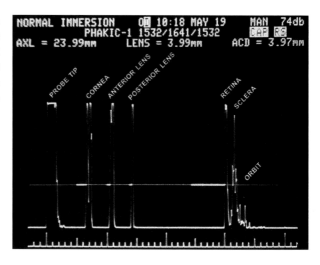

Immersion Echo Pattern Criteria

Looking at the echogram from left to right (Figure 11-4), the following echospikes are noted:

- The first spike is the "main bang" produced by the probe tip.
- The corneal peak should be double spiked and of equal amplitude.
- The anterior lens echo should be 100% tall.
- The posterior lens echo should be as tall as possible depending on the density of the opacity.
- The retinal echo should be 90 degrees from baseline, sharply rising and 100% tall.
- The presence of scleral echo to ensure reading to retina and *not* optic nerve.
- The slow-rising orbital echoes.

Importance of A-Scan Pattern Recognition by Physicians and Technicians Performing Biometry

The key to any biometry exam and intraocular lens (IOL) selection is in understanding the echo patterns, both for the examiner as well as the physician who must evaluate the resultant data and make an educated clinical judgment as to which calculated power to implant. This is paramount to delivering the best possible patient care, having a good postoperative result and ultimately a happy patient, especially in these times of exacting refractive surgery.

The need for pattern recognition has never been greater. In an effort to improve accuracy, we must improve our measurements. It is imperative that technicians performing A-scans be able to recognize a proper echo pattern from one that is inadequate. It is just as imperative that the physician is able to do this, as it is the ophthalmologist who is the deciding factor in these decisions and the final decision must be made from a position of confidence for the best possible patient care.

Physicians and technicians alike should recognize the importance and immense responsibility of performing biometry and IOL calculations. In most states, we as technicians are not legally allowed to write a spectacle prescription, which could be changed tomorrow, but we have the onerous responsibility of measuring an eye and calculating an IOL that would take a second surgery to replace!

This is a huge responsibility and should not be taken lightly either by the technician *or* physician. Ultimately it is the physician's responsibility to review the scan patterns and select the lens choice based on the exam data. However, if the techs haven't been trained to know what a good echo pattern looks like, they may not know they are not getting acceptable scans until the physician review of the A-scan occurs. They then may have to bring the patient back for another A-scan and possibly have to reschedule the surgery until another scan is done, all because the techs were not trained properly or they haven't achieved the necessary skills. Does this sound like A-scan biometry can be delegated to a receptionist or another untrained person?

Ultrasound Measurement of the Challenging Eye

Pier Enrico Gallenga, MD
Adriano Mancini, MD
Anastasia Di Giovanni, MD

The average ultrasound velocity in phakic eyes is about 1550 m/s. It is back-calculated by averaging the measurements taken from the aqueous, vitreous, and the lens. Newer instruments set the correct velocities for aqueous and vitreous at 1532 m/s and for the lens at 1641 m/s or 1620 m/s. A typical echographic diagram of the phakic eye is characterized by a peak derived from the cornea, two peaks from the anterior and the posterior surfaces of the lens, a vitreal acoustic silence, the high retinal peak, and then a medium-high peak from the sclera and low peaks from retrobulbar fat (see Figure 11-4).

This chapter describes the axial length (AL) ultrasonic display and the measurement methods in some challenging eyes that differ from the normal phakic eye; mainly the aphakic eye, pseudophakic eye, the eye with an intumescent cataract, and the eye with silicone-filled vitreous.

The average sound velocity within the ocular structures varies according to the presence (phakic) or absence of a lens (aphakic), the type of cataract, the intraocular lens (IOL) material in a pseudophakic eye, and in the presence of silicone oil in the vitreous cavity (Table 12-1).

The Aphakic Eye

In the aphakic eye, localizing the fovea is more difficult due to the absence of the two lens peaks (in the phakic eye) that help us to direct the ultrasound beam perpendicular to the fovea. An echosignal is detected after the corneal peaks. This signal could be caused by different structures:

- In some cases, it represents the iris surface, especially if maximal midriasis is not obtained. In fact, the sound beam can be larger than the pupil.

TABLE 12-1. Average Sound Velocity in Different Ocular Structures	
Ocular Structures	**Velocity (m/s)**
Aphakia (aqueous and vitreous)	1532
Phakic	1550
Cataract	
nuclear cataract	1610
capsular opacities	1670
nuclear sclerosis + capsular opacities	1641
intumescent cataract	1590
Silicone IOL	980-1090
PMMA IOL	2780
Acrylic IOL	2180
Silicone oil	
silicone oil 1000 centistoke	980
silicone oil 5000 centistoke	1040

- In other cases, it represents the posterior capsule of the lens after an extracapsular cataract extraction.
- And in some cases, this peak could represent the vitreous-aqueous interface after an intracapsular cataract extraction.

Using an immersion technique, the A-scan pattern of an aphakic eye[1] displays the following echospikes from left to right (Figure 12-1):

IS: The initial spike. In an immersion technique, the initial spike can be removed from the screen leaving only the echospikes generated from the eye. In a contact technique, it will be merged with the corneal spike (C).

C: The double-peaked corneal spike.

I: A medium reflective echospike from the iris surface and/or anterior vitreous face.

R: The retinal spike from the anterior surface of the retina.

S: The scleral spike.

O: The orbital spikes.

The initial spike can also be removed from the screen (Figure 12-2), leaving only the echospikes generated from the eye.

The axial length (AL) is measured between the anterior corneal surface (C) and the anterior retinal surface (R) using an average sound velocity of 1.532 mm/μs (1532 m/s), which is the velocity in aqueous and vitreous. Certain units use a slightly higher sound velocity of 1534 m/s to account for the faster speed of sound within the cornea.

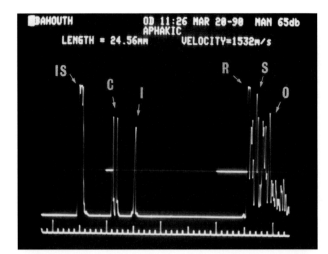

Figure 12-1. A-scan pattern of an aphakic eye, identifying the initial spike (IS), the cornea (C), the iris and/or anterior vitreous face (I), the retina (R), sclera (S), and orbital tissues (O).

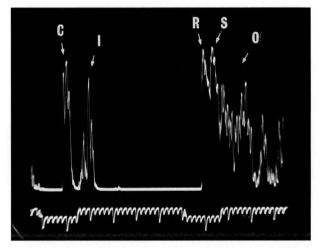

Figure 12-2. A-scan pattern of an aphakic eye where the initial spike has been removed from the screen display, and the only displayed echospikes represent the cornea (C), the iris and/or anterior vitreous face (I), the retina (R), sclera (S), and orbital tissues (O).

In the aphakic eye, the axial length is measured using a sound velocity of 1534 m/s.

If the ultrasound unit uses only fixed 1550 m/s velocity and does not allow the use of 1534 m/s velocity, the AL of the aphakic eye can then be calculated:[1]

Aphakic AL = (1534/1550) x AL measured with 1550 m/s

Figure 12-3. A-scan pattern of a pseudophakic eye identifying part of the initial spike (IS), the cornea (C), the pseudophakos lens (P), the multiple reverberations (M), the retina (R), sclera, and orbital tissues (O).

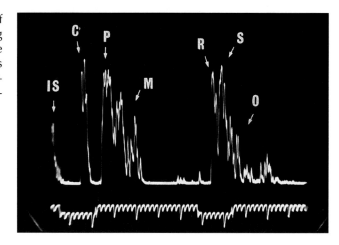

The Pseudophakic Eye

In a pseudophakic eye (Figure 12-3), the A-scan pattern displays the following echospikes:[1,2]

IS: The initial spike. It can be moved out of the screen in an immersion technique. It is merged with the corneal spike (C) in a contact technique.

C: The double-peaked corneal spike.

P: A high-reflective spike from the anterior surface of the pseudophakic lens. It is usually followed by multiple smaller echospikes (M) that represent reverberations of the ultrasound beam between the anterior and posterior surfaces of the implant.

R: The high-reflective retinal spike.

S: The scleral spike.

O: The orbital spikes.

In the pseudophakic eye, we do not have a typical ultrasound tracing because of the ultrasound reverberations causing several artifacts. The operator has to remember to lower the beam's amplification to better differentiate the different peaks and to reduce artifacts.

In the pseudophakic eye, Holladay[3] suggests measuring the AL at the velocity of 1532 m/s and then adding or subtracting a corrected axial length factor (CALF) derived from the different velocity in the lens and in the IOL, rather than using an average velocity:

$$AL = AL_{1532} + CALF$$

Where AL is the true axial length of the pseudophakic eye, CALF is the corrected axial length factor, and AL_{1532} is the axial length measured at the velocity of 1532 m/s.

TABLE 12-2.
Average Sound Velocity and Central Thickness of Currently Available IOLs

Implant	Sound Velocity	Central Thickness
PMMA	2660 m/s	0.6 to 0.8 mm
Silicone	980 m/s	1.2 to 1.5 mm
Glass	6040 m/s	0.3 to 0.4 mm
Acrylic	2200 m/s	0.7 to 0.9 mm

From Shammas HJ. *Intraocular Lens Power Calculations: Avoiding the Errors.* Glendale, CA: New Circle Publishing House; 1996.

To calculate CALF, we need the thickness of the lens (T_L) and the velocity through the lens (V_L). This information is available from the manufacturer; it depends on the IOL power and on the different IOL material.

$$CALF = T_L \times (1 - 1532 / V_L)$$

The final formula will be:

$$AL = AL_{1532} + T_L (1 - 1532 / V_L)^1$$

If the measurement of the AL is taken at a velocity of 1550 m/s, this formula can be converted in:

$$AL_{1532} = (1532/1550) \times AL_{1550}{}^1$$

The average sound velocities (V_L) and central thickness (T_L) of different implants are given in Table 12-2.

MEASURING THE PSEUDOPHAKIC EYE USING THE SPECIFIC CONVERSION FACTOR

The measurement adjustments differ according to the IOL material because the ultrasound velocities vary between silicone, acrylic, and PMMA. Holladay and Prager[4] have described a method for measuring pseudophakic eyes, as follows:

$$TAL = AAL_{1532} + (CF \times T)$$

Where TAL is true axial length, AAL_{1532} is the apparent axial length when using a sound velocity of 1532 m/s, CF is the specific conversion factor, and T is the center thickness of the IOL.

For PMMA, the conversion factor is +0.45, for silicone IOLs, it is -0.56/-0.41 (depending on style and manufacturer), and for acrylic IOLs, it is +0.30. The thickness of the IOL should be obtained directly from the manufacturer.

Acrylic IOLs

Ultrasound velocity through an acrylic IOL is calculated to be 2180 m/s, ie, faster than the 1532 m/s in aqueous. This should cause an apparent AL shorter than real. To avoid this situation, we must calculate a conversion factor. The true AL of an eye with an acrylic IOL is determined by adding the apparent AL measured at 1532 m/s to the conversion factor multiplied by the IOL thickness. We calculate the conversion factor as follows:

$$CF = 1 - (1532/2180) = 0.2972 \sim 0.30$$

$$TAL_{acrylic} = AAL_{1532} + (0.30 \times T)$$

Silicone IOLs

Ultrasound velocity through a silicone IOL is estimated at 980 m/s to 1090 m/s, depending on whether the IOL is a first or second generation IOL. These velocities are slower than the 1532 m/s in aqueous, so that the apparent AL should be longer than real. We can calculate the real AL throughout a conversion factor as follows:

First, we calculate the conversion factor when the AL is performed at 1532 m/s:

For older silicone IOLs, $CF = 1 - (1532/980) = -05633 \sim -0.56$.
For newer silicone IOLs, $CF = 1 - (1532/1090) = -0.4120 \sim -0.41$.

Both of these are negative numbers. So we can conclude:

$$TAL_{silicone} = AAL_{1532} - (0.56 \times T)$$
$$TAL_{silicone} = AAL_{1532} - (0.41 \times T)$$

PMMA IOLs

The ultrasound velocity through a PMMA IOL is estimated at 2780 m/s, so the apparent AL will be shorter. Using the same technique, we can estimate the real AL. We calculate the conversion factor as follows:

$$CF = 1 - (1532 / 2780) = 0.4489 \sim 0.45$$

Then we have the following formula to obtain the true axial length (TAL):

$$TAL_{PMMA} = AAL_{1532} + (0.45 \times T)$$

TABLE 12-3.
IOL Correction Factors

IOL Power (in diopters)	Acrylic IOL (in mm)	PMMA IOL (in mm)	Silicone IOL (in mm)
12-17	+0.20	+0.28	-0.48
18-23	+0.24	+0.32	-0.54
24-27	+0.27	+0.36	-0.65

The correction factors are obtained by the equation (+0.30 x T) for acrylic IOLs, (+0.45 x T) for PMMA IOLs, and (-0.45 x T) for the new silicone IOLs, and where T represents the implant's central thickness.

These values vary slightly depending on the exact dioptric power and size of said implant.

MEASURING THE PSEUDOPHAKIC EYE USING THE IOL CORRECTION FACTORS

When performing A-scan ultrasonography on a pseudophakic eye at a velocity of 1532 m/s, a correction factor should be added to the AL displayed to get a close approximation of the true AL.

The following correction factors have been recommended:[15]
- PMMA IOL: AL_{1532} + 0.4 mm.
- Acrylic IOL: AL_{1532} + 0.2 mm.
- Silicone IOL: AL_{1532} − 0.6 mm for new silicone IOLs and AL_{1532} − 0.8mm for old silicone IOLs.

The correction factors for acrylic and PMMA lenses will be positive and the correction factors for silicone lenses will be negative. Also, the correction factor varies with the IOL's central thickness. For example, this correction factor is as low as +0.14 for a 10 D IOL and as high as +0.25 for a 30 D IOL. The same applies for a PMMA IOL (+0.20 to +0.60) and for a silicone IOL (-0.50 to -0.70) (Table 12-3).

MEASURING THE PSEUDOPHAKIC EYE USING AN AVERAGE SOUND VELOCITY

If the eye is to be measured with an average sound velocity instead of using preceding formulas, the following velocities are recommended:[1]
- 1555 m/s for an eye with a PMMA IOL.
- 1476 m/s for an eye with a silicone IOL.
- 1549 m/s for an eye with a glass IOL.
- 1554 m/s for an eye with an acrylic IOL.

If a pseudophakic eye is measured at the phakic average velocity of 1550 m/s, the error is less than 0.1 mm for the eye with a PMMA, glass, or acrylic IOL. However, this error exceeds 1.0 mm for the eye with a silicone IOL.

Figure 12-5. A-scan pattern of an eye with silicone-filled vitreous where the vitreous cavity is measured at 1532 m/s requiring separate measurements of the different ocular compartments, recalculation of the correct vitreous cavity's depth and of the true axial length. (Courtesy of Rhonda G. Waldron, MMSc, COMT, CRA, ROUB, RDMS.)

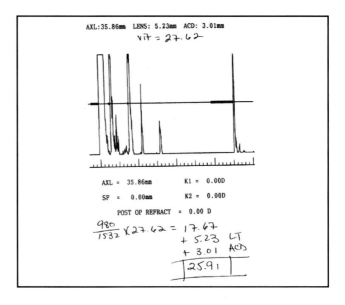

Figure 12-6. A-scan pattern of an eye with silicone-filled vitreous measured with the Accusonic unit and where the vitreous cavity is measured at 980 m/s. (Courtesy of Rhonda G. Waldron, MMSc, COMT, CRA, ROUB, RDMS.)

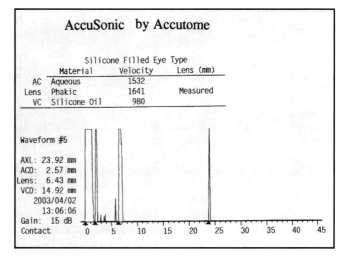

ed to have higher changes in refraction and AL than the eyes filled with 1000 cSt silicone oil.[8] The 1000 cSt oil has a velocity of 980 m/s, whereas the 5000 cSt oil's velocity is 1040 m/s. The low velocity within the silicone oil will cause an erroneous measurement of the vitreous cavity depth (VCD). The formula to correct the AL in any silicone oil-filled vitreous is:

1. $VCD_{1532} = AL - (ACD + LENS)$
2. $VCD_{corrected} = VCD_{1532} \times (1/1532) \times 980 \text{ m/s }$*
3. $AL_{Corrected} = VCD_{corrected} + ACD + LENS$

In some cases, silicone oil has to remain for a long period in the vitreous cavity. In this case we have to consider some IOL power adjustments. In these patients,

---

*or 1040 m/s (depending on the oil placed in the patient's eye)

PMMA lenses are preferred because silicone oil alters the refractive power of the posterior surface of the IOL.

Patel in 1995 and then Meldrum suggest using the following correcting formula to calculate the additional power to add to IOL power calculation in case of silicone filled vitreous:

$$\text{Additional IOL power} = \{(Ns - Nv) / (AL - ACD)\} \times 1000 *$$

Where Ns is the refractive index of silicone oil (1.4034), Nv is the refractive index of vitreous (1.336), AL is the axial length in millimeters, and ACD is the anterior chamber depth in millimeters.

The additional IOL power for a silicone oil-filled vitreous is +3.0 to 3.5 D to obtain emmetropia.

In case of cataracts not so advanced or mature to impair fundus examination, partial coherence interferometry (Zeiss IOLMaster [Zeiss Humphrey Systems, Dublin CA]) is preferred over ultrasonography to calculate the AL of the human eye filled with silicone oil.

In cases of eyes filled with gas or perfluorocarbonates, ultrasound echoes are blocked. Moreover, we can have two-media tamponade or three-media tamponade. In these cases, it is advised to position the patient's head temporally in order to move the media and to have a chance to measure the AL. This is possible because the vitreous cavity is not completely filled with these media.

References

1. Shammas HJ. *Intraocular Lens Power Calculations: Avoiding the Errors.* Glendale, CA: New Circle Publishing House; 1996.
2. Gallenga PE, Cennamo G. *Indicazioni dell'esame ecografico nella pseudofachia.* Milano: CIICA (Italian Intraocular Implant Club), Edizioni Medicina Internazionale. 1985;73-82.
3. Holladay JT. Standardizing constants for ultrasonic biometry, keratometry and intraocular lens power calculations. *J Cataract Refract Surg.* 1997;23:1356-1370.
4. Holladay JT, Prager TC. Accurate ultrasonic biometry in pseudophakia (letters). *Am J Ophthal.* 1992;115(4):536-537.
5. Bellow JC. *Cataract and Anomalies of the Lens: Growth, Structure, Composition, Metabolism, Development, Growth Disorders and Treatment of the Crystalline Lens.* St. Louis, MO: CV Mosby; 1944.
6. Pallikaris I, Gruber H. Determination of sound velocity in different forms of cataracts. *Documenta Ophthalmologica.* 1981;29:165-169.
7. Gallenga PE, Del Duca M. Study in silicone oil in vitreoretinal surgery. In: Ossoinig KC, ed. *Ophthalmic Echography SIDUO X.* Dr. W. Junk Publishers; 1987:48.
8. Seo MS, Lim ST, Kim HD, Park BI. Changes in refraction and axial length according to the viscosity of intraocular silicone oil. *Korean J Ophtal.* 1999;13(1):25-29.

---

*or 5000, depending on the type of silicone oil used.

Avoiding Errors in Axial Length Measurement

In an accurate axial length (AL) measurement, the technician aligns the ultrasound beam with the optical axis of the eye by being perpendicular to the four major surfaces of the eye: the anterior surface of the cornea, anterior surface of the lens, the posterior surface of the lens, and the anterior surface of the retina. Biometry units that are not equipped with an oscilloscope or a screen that displays the actual scan have a high error rate and are definitely not recommended.

Biometry units that are not equipped with an oscilloscope or a screen that displays the actual scan have a high error rate and are definitely not recommended.

Errors in an AL measurement are due to an improper technique yielding shorter or longer measurements.[1-3] Often, manufacturers recommend using the average value of multiple measurements to improve precision and avoid errors. Although this is a good practice, one should remember that multiple readings of an erroneous measurement will still yield an erroneous average measurement. The best technique is the one that provides both precision and accuracy. Precision allows the reproducibility of multiple measurements and a decrease in the standard deviation and standard error, while accuracy provides a result close to the true value. Accuracy requires calibration with a known standard.

Errors in AL measurement rarely occur with high precision A-scan biometry using an immersion technique. Such errors are more commonly noted with contact A-mode biometry, mainly due to corneal compression. Other causes of error include a faulty technique with failure to correctly display the ocular structures and the presence of intraocular pathology.

Corneal Compression

Corneal compression[4,5] is the most common cause of shorter AL measurements. It occurs with contact A-scan biometry where the ultrasound probe is in touch with the cornea. An unskilled technician can indent the cornea with the A-scan probe more than needed, resulting in a shallower anterior chamber depth (ACD) and a shorter AL, even though an acceptable A-scan echogram has been displayed on the screen.

Corneal compression is more pronounced with hand-held probes where there is no chin rest to stabilize the patient's head, and with large, flat surfaced probes that have the tendency to applanate a larger surface of the cornea. Corneal compression is less accentuated with the smaller probes that have a concave surface to mold the anterior corneal surface. No corneal compression occurs with immersion A-scan biometry where the ultrasound probe remains at a certain distance from the cornea.

Entering shorter measurement of the AL in intraocular lens (IOL) power calculations will call for the use of a stronger IOL than is actually required, resulting in an induced myopia in the final postoperative refraction.

Clinical Application

Figures 13-1A and B are taken from the same eye. Although both scans look acceptable, the AL is 0.3 mm shorter in Figure 13-1A, where the technician has compressed the cornea during the examination.

Off-Axis Measurement

Off-axis measurement[6,7] occurs when the ultrasound beam is not perpendicular to the surfaces of the eye.

SMALL OR ABSENT POSTERIOR LENS SPIKE

A minimal off-axis scan is characterized by the absence of the posterior lens spike or the presence of a very small one. The remaining echospikes from the cornea, anterior surface of the lens, and the retina usually appear normal.

Clinical Application

Figures 13-2A and B are taken with the contact technique from the same eye. Note the small posterior lens spike in Figure 13-2A due to a slight off-axis ultrasound direction and resulting in a 0.14 mm shorter measurement when compared with the ultrasound pattern in Figure 13-2B.

POORLY IDENTIFIED RETINAL SPIKE

A larger off-axis measurement occurs when the patient is not looking at the fixation light due to the presence of a dense cataract or to the inability of the patient to hold the eye in a steady position. A larger off-axis scan is characterized by the

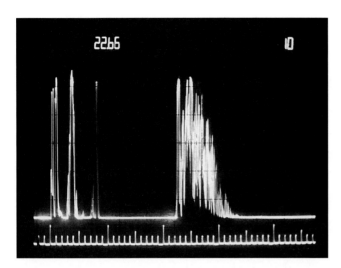

Figure 13-1A. A-scan display of a phakic eye with corneal compression showing an axial length reading of 22.65 mm.

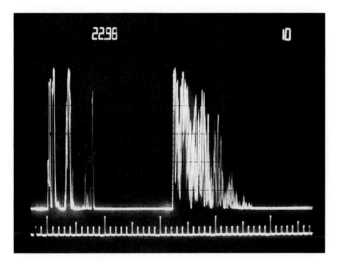

Figure 13-1B. A-scan display of the same eye with no corneal compression showing an axial length reading of 22.96 mm.

absence of the posterior lens spike and the presence of a poorly identified or jagged retinal spike. The biometer will miss the retinal spike and read a longer measurement between the corneal and the scleral spike.[8]

Clinical Application

Figures 13-3A and B are taken with the contact technique from the same eye. Note the small lens spikes and the jagged retinal spike in Figure 13-3A due to a large off-axis ultrasound direction, resulting in a 0.51 mm shorter measurement when compared with the ultrasound pattern in Figure 13-3B.

Figures 13-4A and B are taken from the same eye. In Figure 13-4A, the retinal spike is weak and the measurement is instead made to the sclera, yielding a false measurement of 25.73 mm. In Figure 13-4B, the retinal spike is sharp and steeply rising with a correct measurement of 24.07 mm.

Figure 13-2A. A-scan display of a phakic eye with a slight off-axis measurement of 23.19 mm. Note the presence of the small posterior lens surface echospike (arrow).

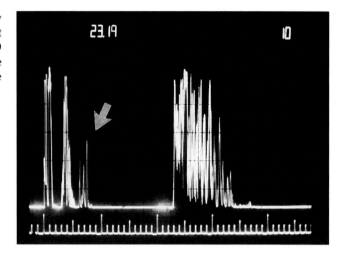

Figure 13-2B. A-scan display of the same eye with an on-axis measurement of 23.33 mm.

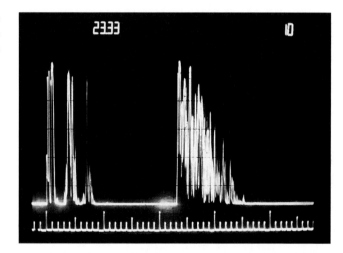

Figure 13-3A. A-scan display of a phakic eye with a large off-axis measurement of 21.87 mm. Note the presence of the small lens spikes (L1 & L2) and the jagged retinal spike (R).

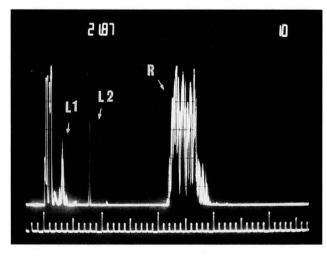

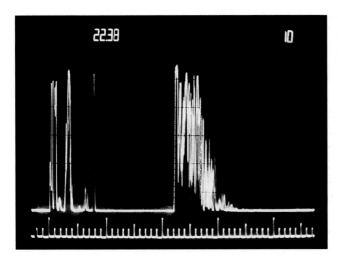

Figure 13-3B. A-scan display of the same eye with an on-axis measurement of 22.38 mm.

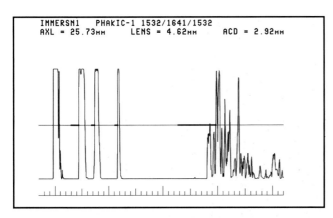

Figure 13-4A. A-scan display of a phakic eye with a poorly defined retinal spike. The erroneously long measurement of 25.73 mm is taken between the cornea and the sclera (instead of the retina).

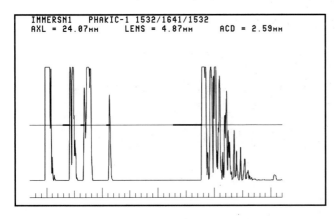

Figure 13-4B. A-scan display of the same eye with a sharp and steeply rising retinal spike. A correct measurement of 24.07 mm is obtained.

Figure 13-5A. A-scan display of a phakic eye taken with the I³ unit. The unit recognizes the absence of the posterior lens echospike and does not give any measurement.

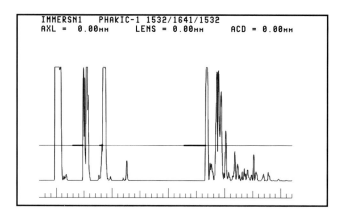

Figure 13-5B. A-scan display of the same eye with a well-identified posterior lens spike and an accurate axial length measurement.

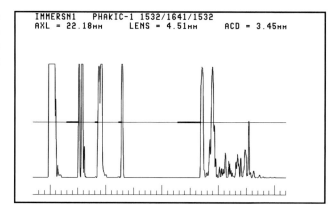

A proper technique and the identification of the ocular structures on the A-scan will prevent any off-axis measurement. New biometry units such as the Innovative Imaging Inc. (I³) biometer (Innovative Imaging Inc., Sacramento, CA) are equipped with an internal mechanism that prevents a measurement reading if the unit does not recognize the echospikes generated from the cornea, anterior lens surface, posterior lens surface (Figures 13-5 A and B), retina, and sclera (Figures 13-6 A and B).

Incorrect Placement of the Electronic Gates

Some ultrasound units allow total movements of the gates. The wrong placement of the electronic gates can read the measurement between two unrelated surfaces instead of between the cornea and the retina.

Clinical Application

Figures 13-7A and B are taken with a contact technique from the same eye. In Figure 13-7A, the corneal gate was inadvertently moved to the right, resulting in a measurement of 19.98 mm in this emmetropic eye. This measurement was taken between the anterior surface of the lens and the anterior surface of the retina.

In Figure 13-7B, the corneal gate was moved back to its original position and the correct measurement of 22.88 mm was obtained.

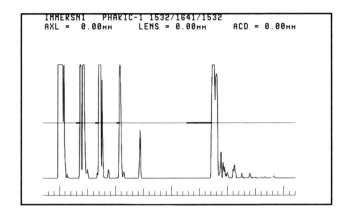

Figure 13-6A. A-scan display of a phakic eye taken with the I[3] unit. The unit recognizes the absence of the scleral spike and does not give any measurement.

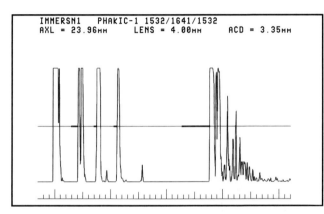

Figure 13-6B. A-scan display of the same eye with well-identified retinal and scleral spikes and an accurate axial length measurement.

Intraocular Pathology

In presence of vitreous pathology, the A-scan ultrasound pattern can sometimes be difficult to interpret. In such cases, a B-scan and/or standardized A-scan makes the correct diagnosis.[9-12] The surgeon will then decide whether the cataract surgery is indicated or not, and whether an intraocular implant should be used if the cataract is removed. In most cases, the pathology is uniocular, and cannot be visualized due to the mature or almost mature cataract.

PRESENCE OF ASTEROID HYALOSIS

The presence of asteroid hyalosis[9] will create echospikes within the vitreous cavity that can be confused by the biometer as the posterior lens surface or the retinal surface. By decreasing the biometer's system sensitivity, the amplitude of all the echospikes will decrease to a point where the weaker vitreous spikes almost disappear. Also, when in doubt as to the nature of the vitreous pathology, a B-scan ultrasound can be helpful (Figure 13-8).

Figure 13-7A. A-scan display of a phakic eye where the overlight corneal gate is inadvertently moved to the right, yielding a false measurement of 19.98 mm from the anterior lens surface to the retina.

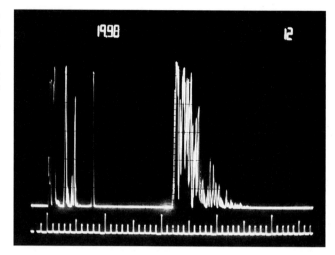

Figure 13-7B. A-scan display of the same eye where the overlight corneal gate has been repositioned over the corneal spike, yielding a correct measurement of 22.88 mm.

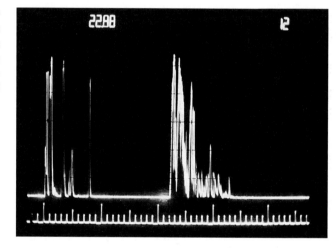

Figure 13-8. B-scan display of an eye with asteroid hyalosis. Note the presence of multiple small echogenic opacities within the vitreous cavity. These opacities are very dense and highly reflective.

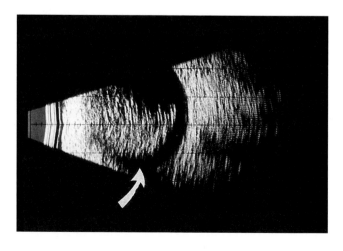

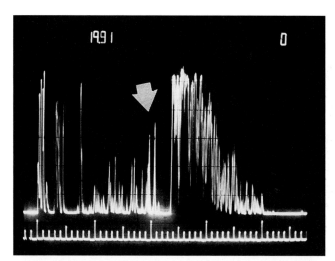

Figure 13-9A. A-scan display of a phakic eye with asteroid hyalosis. The large overlight retinal gate mistook the posterior asteroid bodies (arrow) for the retina, yielding a false measurement of 19.91 mm.

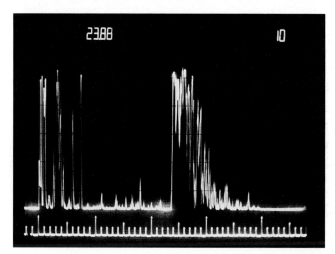

Figure 13-9B. A-scan display of the same eye with decreased system sensitivity and a repositioned retinal gate, yielding a correct measurement of 23.88 mm.

Clinical Application

Figures 13-9A and B are taken with a contact technique from the same eye with asteroid hyalosis. In Figure 13-9A, the retinal gate picked up the presence of the calcified bodies within the vitreous. This resulted in a measurement of 19.91 mm.

In Figure 13-9B, the system sensitivity was decreased by 10 dB and the retinal gate was repositioned to avoid the vitreous pathology, resulting in the correct measurement of 23.88 mm.

Figure 13-10A. A-scan display of a phakic eye with asteroid hyalosis. The unit mistook an anteriorly located asteroid body for the posterior lens surface, yielding a false lens thickness of 9.28 mm and a false axial length measurement of 25.56 mm.

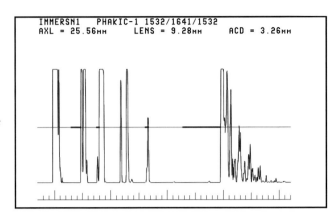

Figure 13-10B. A-scan display of the same eye where the posterior lens surface is correctly identified, yielding a correct lens thickness of 5.44 mm and a correct axial length of 25.36 mm.

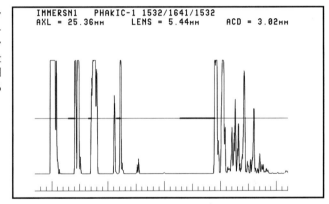

Clinical Application

Figures 13-10A and B are taken with an immersion technique from the same eye with asteroid hyalosis. In Figure 13-10A, the anteriorly located calcified bodies within the vitreous are mistaken for the posterior lens spike, giving a false lens thickness measurement of 9.28 mm and a wrong AL measurement of 25.56 mm.

In Figure 13-10B, the posterior lens spike is correctly identified, yielding a lens thickness measurement of 5.44 mm and a correct AL measurement of 25.36 mm.

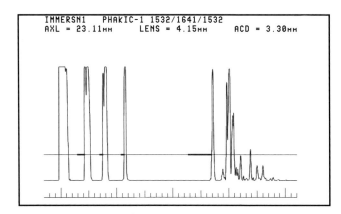

Figure 13-11A. A-scan display of a phakic eye with retinal detachment where the retinal spike is separated from the posterior layers of the eye.

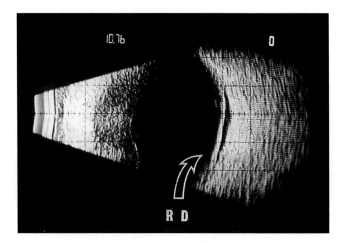

Figure 13-11B. B-scan display of the detached retina (RD).

RETINAL DETACHMENT

In an uncomplicated retinal detachment, the retina is totally separated from the pigment epithelium, but remains attached to the optic disc and to the ora serrata.

Clinical Application

This 72-year-old lady presents with decreased vision in both eyes, more so in the right eye for the past 3 years. Visual acuity is light perception with a mature cataract. Measurement of the AL was very difficult to perform (Figure 13-11A) and showed the presence of an extra spike in front of the posterior segment layers. The B-scan ultrasound revealed a total retinal detachment (Figure 13-11B).

Figure 13-12. B-scan display of a phakic eye with disciform macular degeneration showing the flat echogenic subretinal elevation caused by bleeding and scarring.

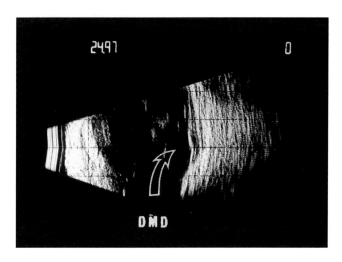

MACULAR PATHOLOGY

A disciform macular degeneration (DMD) is caused by subretinal scar tissue formation secondary to bleeding. The choroid is elevated, and in certain cases it can simulate a choroidal tumor.

Clinical Application

This 65-year-old lady presented with an immature cataract in the left eye with nuclear sclerosis and very dense posterior subcapsular precipitates, making fundus examination impossible. The fellow eye had 20/40 vision, minimal cataract changes, and few macular drusens.

Measurement of the AL was difficult and yielded a measurement of 20.85 mm, while the other eye measured 23.20 mm. B-scan revealed the presence of macular degeneration, subretinal hemorrhage, and a diffuse vitreous hemorrhage (Figure 13-12).

POSTERIOR POLE STAPHYLOMA

A posterior pole staphyloma is usually present in severe axial myopia. Occasionally it can be undetected if associated with a mature cataract in an eye with unilateral axial myopia.[10]

Clinical Application

This 48-year-old male is referred by his ophthalmologist for repeat AL measurement and IOL calculations because of difficulties in measurement in his office. An immersion technique was used and yielded a measurement of 26.60 mm in the cataractous eye (Figure 13-13A). B-scan examination shows the increased curvature of the posterior pole in the cataractous eye (Figure 13-13B) with a central staphyloma.

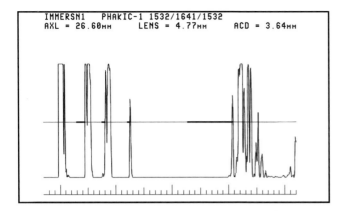

Figure 13-13A. A-scan display of a phakic eye with a posterior pole staphyloma. Note the presence of a weak retinal spike due to the increased curvature of the posterior pole.

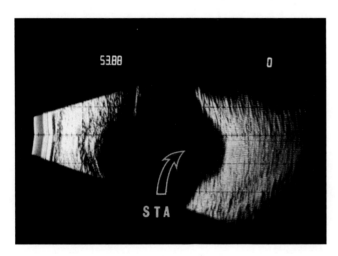

Figure 13-13B. B-scan display of an eye with a posterior pole staphyloma (STA).

Recommendations

There is no fool-proof method to avoid an error in AL measurement, but I would like to share a protocol that has proven to be very effective.[5]

1. Bilateral AL measurements are performed by a certified technician or by the ophthalmologist. High precision immersion A-scan biometry is strongly recommended for reproducible measurements. Each eye is measured three times and pictures or print-outs are obtained. Consistency is very important. Variations in the ultrasound probe design, in the sound velocity, or in the technique can cause up to 0.5 mm difference in measurement.

2. The technician reviews the A-scans to ensure that the ocular structures are properly identified and that there are no errors due to an off-axis examination.

3. Prior to surgery, the surgeon personally reviews the pictures of the measurements. The ultrasound results are correlated with the clinical data and an implant is chosen. In most cases, both eyes are within 0.5 mm of each other and the ultrasound results correlate with the clinical data; hypermetropes usually

measure less than 23 mm, emmetropes between 23 mm and 24 mm, and true myopes over 24 mm. However, discrepancies will be encountered in some cases. The surgeon should then use his or her clinical judgment in analyzing the data. If the surgeon is not satisfied with the calculations or suspects a possible error, he or she should have the measurements repeated, either under his or her supervision, or by an outside consultant. When needed, a B-scan examination will rule out intraocular pathology.

References

1. Ossoinig KC, Byrne SF, Weyer NJ. Standardized echography, Part II: performance of standardized echography by the technician. *Int Ophthalmol Clin*. 1979;19:283-285.
2. Shammas HJ. Manual versus electronic measurement of the axial length. In: *Ultrasonography in Ophthalmology, Proceedings of the 1982 Ninth SIDUO Congress*. Hillman JS, LeMay MM, eds. The Hague: Dr. W. Junk Publishers; 1983;225-229.
3. Salz JJ, Reader AL III. Lens implant exchanges for incorrect power: results of an informal survey. *J Cataract Refract Surg*. 1988;14:221-224.
4. Binkhorst RD. The accuracy of ultrasonic measurement of the axial length of the eye. *Ophthalmic Surgery*. 1981;12:363-365.
5. Shammas HJ. *Atlas of Ophthalmic Ultrasonography and Biometry*. St. Louis: CV Mosby Co; 1984.
6. Bonnac JP, Massin M, Poujol J. Comparison between the measures of length of a myopic eye in the axis and in the para-axial regions. In: Francois J, Goes F, eds. *Ultrasonography in Ophthalmology*. Basel, Switzerland: S. Karger; 1975.
7. Shammas HJ. Ultrasound diagnosis in Unilateral Axial Myopia. In: *Ophthalmic Ultrasonography. Proceedings of the 12th SIDUO Congress*. Sampaolesi R, ed. Dordrecht, The Netherlands: Kluwer Academic Publishers. Documenta Ophthalmologica Series. 1990;149-153.
8. Ossoinig KC. Standardized echography: basic principles, clinical applications, and results. *Int Ophthalmol Clin*. 1979;19:127-210.
9. Martin RG, Safir A. Asteroid hyalosis affecting the choice of intraocular lens implant. *J Cataract Refract Surg*. 1987;13:62-64.
10. Shammas HJ, Milkie CF. Mature cataracts in eyes with unilateral axial myopia. *J Cataract Refract Surg*. 1989;15:308-311.
11. Tjan TT. Ultrasonic diagnosis of ocular pathology: its significance for cataract surgery with and without IOL implantation. *American Intra-Ocular Implant Society Journal*. 1979;5:15-21.
12. Wainstock MA. Ultrasonography: its role in the success of intraocular implant surgery. *Int Ophthalmol Clin*. 1979;19:43-50.

Optical Biometry Using Partial Coherence Interferometry

Wolfgang Haigis, PhD

One and a half centuries after ultrasound was found to help bats "see with their ears", Mundt & Hughes[1] were the first to apply echography to record a human eye length in 1956. Since then, ultrasound biometry has become established as a precise and reliable tool for the measurement of ocular distances and is applied particularly to the measurement of axial length (AL) and its segments for the calculation of intraocular lenses (IOL).

In 1999, Carl Zeiss of Jena, Germany, introduced a new non-contact biometry instrument based on optical methods: the IOLMaster (Zeiss Humphrey Systems, Dublin, CA). This device is an all-in-one instrument allowing all measurements and calculations necessary for proper IOL power selection in cataract and refractive surgery. Its measurement options (Figure 14-1) include: determination of AL and anterior chamber depth (ACD), determination of corneal curvatures (or powers), of the (horizontal) corneal white-to-white diameter, and the position of the visual axis. An outstanding feature of the IOLMaster is its AL measurement module, which is based on *partial coherence interferometry* (PCI), also termed *laser Doppler interferometry* (LDI). Applying this technique in optical biometry is known as *laser interference biometry* (LIB), or *optical coherence biometry* (OCB). In the last few years, optical biometry has gained much popularity and has established itself as a possible alternative to classic A-scan ultrasound.

The use of this method for eye length measurement[2-4] goes back to Fercher et al.[5] In its tomographic variant as optical coherence tomography (OCT), the technique has found widespread use in ophthalmology as well.

Our laboratory (Biometry Department of the University Eye Hospital, Wuerzburg) has been involved in the development of the IOLMaster since 1997.[6-12]

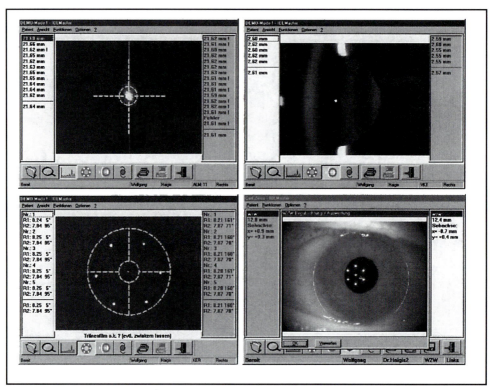

Figure 14-1. Measurement features of the Zeiss IOLMaster: determination of axial length (top left), anterior chamber depth (top right), corneal curvature/power (bottom left), white-to-white diameter and position of visual axis (bottom right).

Dual Beam Partial Coherence Interferometry

The principle behind PCI biometry is a dual beam technique using wavelets of infrared light (wavelength λ = 780 nm). A schematic sketch of the optical setup is depicted in Figure 14-2.[8,13] Two coaxial beams I_m and I_f are created in a Michelson interferometer by a laser diode (LD) with short coherence length c ($c \approx 130 \ \mu m$), a beam splitter (BS1), a fixed mirror (M_f), and a moving (measurement) mirror (M_m). These beams propagate into the eye, and are reflected at the different ocular interfaces, especially at the cornea (C) and the retina (R). Returning from the eye through another beam splitter (BS2), all reflected partial beams are directed onto a photo detector (PD). If two wavelets meet each other within the (short) coherence length c of the laser diode, ie, 2OPL-2d $\leq c$, they can interfere and their intensity distribution is sensed by the detector. This is the case for wavelets $I_f(R)$ from the retina and $I_m(C)$ from the cornea if the displacement (d) of the moving mirror (M_m) is equal to the optical path length (OPL) through the eye. The high precision of this measurement technique stems from the accuracy with which the displacement of M_m can be determined.

It is evident that longitudinal eye movements do not affect the measurement, whereas transversal displacements can make it impossible to acquire a good quality signal.

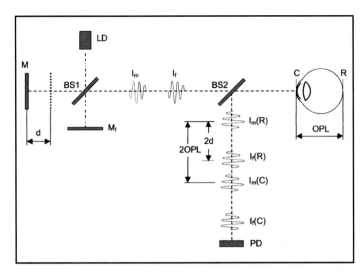

Figure 14-2. Principle layout for optical coherence biometry:[8,13] two coaxial beams I_m and I_f are created in a Michelson interferometer and reflected at different ocular interfaces. If two components, eg, $I_f(R)$ and $I_m(C)$, of the two beams are close enough (if 2OPL-2d $\leq$ c; c: coherence length), they can interfere and their intensity distribution is sensed at the photodetector (PD).

Figure 14-3 shows an optical A-scan obtained from a glass sphere as the IOLMaster internally sees it. There are basically two strong reflections: one from the anterior (corneal), the other one from the posterior (retinal) side. Apart from the two main signals, each one of them is surrounded by (unwanted) additional periodical side lobes that are due to the laser diode's characteristics (periodicity ≈ 0.85 mm). Since the corneal peak will always be at the beginning of the optical A-scan, it is not displayed on the IOLMaster's screen; instead, only the signal of interest, ie, the one from the ocular back wall, is shown (Figure 14-4). As can be seen in Figure 14-4, which is a zoomed IOLMaster display of an eye with a 22.97 mm AL, there is often more than one peak from the posterior ocular interfaces. The main signal usually originates from the retinal pigment epithelium (RPE).[2] Often, a smaller signal (typically 0.2 to 0.3 mm anterior to the main peak) can be observed, which stems from the internal limiting membrane (ILM). Sometimes, choroidal structures following the RPE signal may also show up. Choroidal signals are slightly broader than RPE signals and may consist of two unresolved peaks from different choroidal layers. Usually,[14] they are found some 0.15 to 0.25 mm posterior to the retinal pigment epithelium. Signal #4 in Figure 14-4 is some 0.5 mm away from the RPE and some 0.8 mm from the ILM. It may be due to a deep choroidal vessel, or—more likely—a side lobe to the ILM signal.

One advantage of optical biometry is that the AL measurement is performed along the visual axis since the patient is asked to fixate onto the laser spot. The visual axis, however, does not intersect the lens surfaces at right angles. Consequently, it is not possible to detect lenticular reflections along the visual axis;[15] segmental measurements like in ultrasound are not possible along this axis. To overcome these limitations, the optical axis has to be used, ie, fixation onto the laser spot has to be eliminated. Instead, the eye to be measured has to be made to fixate onto a target away from the measurement axis by some 5 degrees.

One advantage of optical biometry is that the AL measurement is performed along the visual axis since the patient is asked to fixate onto the laser spot.

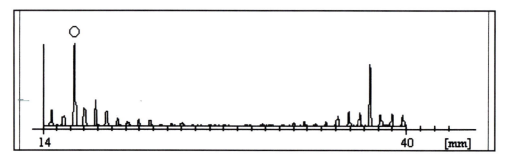

Figure 14-3. Optical A-scan, measured on a phantom (glass sphere, L=20.63 mm). The front surface reflection (stemming from the cornea in the case of an eye)—unlike here—is not displayed on the IOLMaster screen, only the back surface reflection (from the retinal pigment epithelium in an eye).

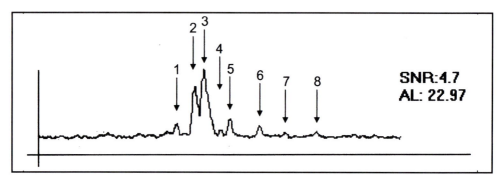

Figure 14-4. Optical A-scan of ocular back wall. Main signal #3 stemming from the retinal pigment epithelium (RPE) shows side lobes (#1, 5, 6, 7, 8) due to the characteristics of the laser diode. The periodicity of these unwanted artifacts is 0.85 mm. Signal #2 originates from the inner limiting membrane (ILM) with a distance ILM-RPE of 0.28 mm. Signal #4 is 0.50 mm away from RPE and 0.78 mm from ILM. It may be due to a choroidal vessel or—more likely—a side lobe of the ILM signal.

Performing these measurements is not too difficult provided the reflections from the lens are not too weak. This is the case in pseudophakic eyes.

Figure 14-5 shows an example of such an optical A-scan of a pseudophakic eye, measured at λ = 850 nm along the *optical* eye axis. Similar measurements, yet only of the anterior ocular segment, have been described by Findl et al.[4,16] The scan of Figure 14-5 shows reflections of all ocular interfaces and is equivalent to a segmental ultrasound measurement.

Optical Path Length and Geometrical Distance

Figure 14-5 was not obtained with the standard version of the Zeiss IOLMaster, but with a modified one. The hardware of the current instrument does not allow segmental measurements. The primary measurement parameter is the optical path

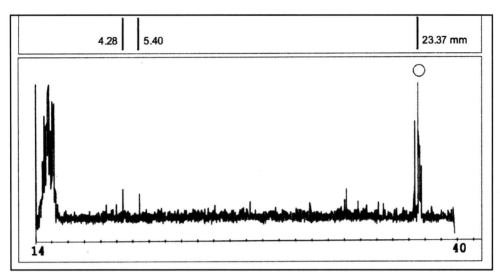

Figure 14-5. Optical A-scan of a pseudophakic eye, measured at λ = 850 nm along the optical axis of the eye. The maximum peak at the beginning of the signal trace stems from the anterior cornea, the following group from the posterior cornea. Two distinct peaks representing the IOL interfaces (4.28 mm, 5.40) are nicely displayed. The last peak with maximum amplitude at 23.37 mm represents the retinal pigment epithelium.

length (OPL) between the anterior corneal surface and the retinal pigment epithelium (measured along the visual axis). The geometric distance AL_{PCI} between these two surfaces is given by:

$$AL_{PCI} = OPL/n_{PCI}$$

Where $n_{PCI} = 1.3549^2$ is the mean group[17] refractive index of the eye.

Using a mean refractive index in laser biometry is directly equivalent to using a mean sound velocity in acoustical biometry.

Clinical Application

Figure 14-6 shows a comparison of axial lengths AL_{PCI} measured with a prototype (ALM) of the IOLMaster with the respective values AL_{OCU} obtained in contact coupling technique with an OcuScan (Alcon, Irvine, CA). IOLMaster ALs were longer by 0.32 ± 0.21 mm (range: -0.41 mm to +0.91 mm). This was to be expected, since ultrasound measures up to the internal limiting membrane. PCI biometry, however, measures up to the retinal pigment epithelium. Axial lengths by both methods correlate very well (R = 0.991) as can be seen from Figure 14-6. The slope (0.942) of the regression line is smaller than 1, caused by the refractive index in the lens, which itself is over-represented in short eyes as compared to long eyes.

Figure 14-6. Optical axial lengths (AL_{PCI}) measured with the ALM prototype of the Zeiss IOLMaster versus axial lengths (n = 134) obtained with the Alcon OcuScan (AL_{OCU}).

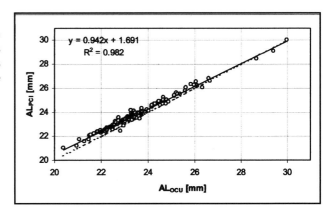

IOLMaster Calibration

The comparisons between acoustical and optical biometry have shown that both methods measure different distances in different directions.[8,10] Since clinical experience in IOL calculation and cataract and refractive surgery has been based on ultrasound examinations, it was decided to calibrate the IOLMaster AL measurements against the most precise available ultrasound immersion data. For this purpose, more than 600 eyes were examined optically and acoustically[8] using the Grieshaber Biometric System (GBS) (Alcon/Grieshaber, Fort Worth, TX), which allows high precision simultaneous segmental immersion measurements with a spatial resolution of 22 µm and a reproducibility of 22 ± 24 µm. A calibration curve was deduced, which was subsequently wired into the market version of the IOLMaster. Thus, the IOLMaster behaves as if it were an extremely precise ultrasound immersion instrument; its AL readouts are directly comparable to results obtained ultrasonically in immersion technique.

Although it is not possible for the IOLMaster to apply laser biometry for segmental measurements, the instrument nevertheless offers a measurement option for the determination of the ACD. This is accomplished by evaluating a slit image of the anterior segment of the eye (see Figure 14-1 top right) using sophisticated image analysis software. Again, the resulting ACDs are calibrated against precise immersion ultrasound values. This was carried out on the basis of more than 800 comparative measurements performed in our laboratory. IOLMaster ACDs, therefore, are equivalent to immersion ACDs. As such they are likely to be a little longer (by 0.1 to 0.2 mm) than ACDs derived from contact ultrasound measurements. Also, similar to an ultrasound ACD, an IOLMaster ACD is defined as the distance from the anterior surface of the cornea to the anterior surface of the lens.

ADDITIONAL MEASUREMENT FEATURES OF THE IOLMASTER

The IOLMaster is designed to allow all measurements necessary for IOL calculation to be performed with one instrument. Apart from biometry, the knowledge of corneal curvatures or corneal refractive powers is mandatory for the proper selection of IOLs.

Keratometry is performed classically by evaluating the Purkinje images of suitable luminescent objects. For this purpose, six infrared light-emitting diodes (LED)

TABLE 14-1.

Inter- and Intra-Observer Variabilities for Measurements of Axial Length, Anterior Chamber Depth, and Corneal Radius with the Zeiss IOLMaster

		Axial Length	ACD	Corneal Radius
Intra-observer variability	[μm]	10.9	31.9	11.3
Coefficient of variation	[%]	0.04	0.94	0.15
Inter-observer variability	[μm]	11.8	37.7	13.4
Coefficient of variation	[%]	0.05	1.12	0.17
Reliability	[%]	100.0	97.8	99.6

Data published in part in Haigis,[9] else unpublished.

arranged in a hexagonal pattern are used (see Figure 14-1 bottom left). Their images on the cornea are digitized by the IOLMaster's system CCD-camera. Corneal curvatures are derived from the positions of the individual LED images.

Images grabbed by the system camera are also used to derive the (horizontal) corneal white-to-white (WTW) diameter (see Figure 14-1 bottom right). Furthermore, the position of the visual axis can be determined by evaluating the corneal image of the IOLMaster's fixation LED. Both measurement features—WTW and visual axis position—require no hardware change and are available as software updates.

Reproducibility, Observer Dependence, Comparison with Other Techniques

Defining reproducibility as the average standard deviation of five consecutive measurements, we obtained an AL reproducibility of 23 ± 15 μm for the IOLMaster and 22 ± 24 μm for our high precision immersion measurements with the GBS.[8,18,19]

In another study (mostly unpublished) involving measurements of AL, ACD, and corneal radii of 29 volunteers, repeated at three different times by four examiners (two experienced, two beginners), we determined the inter- and intra-observer variability of the respective measurement parameters for the IOLMaster. Table 14-1 lists the results. The intra-observer variability for AL, ACD, and corneal radius of curvature were 10.9 μm, 31.9 μm, and 11.3 μm. The respective results for the inter-observer variability were 11.8 μm, 37.7 μm, and 13.4 μm. The reliabilities were 100.0% for AL, 97.8% for ACD, and 99.6% for corneal radius. These results indicate that IOLMaster measurements are extremely user-independent. Comparable results were also obtained by Vogel et al.[20,21]

In a number of studies, different IOLMaster individuals were rechecked against our high precision GBS system. In one of them,[9] a correlation coefficient of 98.8% and a regression equation of $AL_{IOLMaster} = 1.0006 \times AL_{GBS} + 0.0337$ was found for the correlation between IOLMaster and GBS ALs. Another one showed a mean difference $AL_{IOLMaster} - AL_{GBS}$ of -10 ± 19 µm (range: -770 to +420 µm, median: 10 µm; n = 146).[19] In the same study, the respective values of the ACD as measured with the two instruments were compared, resulting in a mean difference of 0.03 ± 0.18 mm (range: -0.40 to +0.68 mm; median 0.00 mm; n = 151). Also, the IOLMaster keratometry module was compared to an Alcon (Renaissance Series) handheld keratometer with a mean difference (IOLMaster-Alcon) of the averaged corneal radii of –10 ± 50 µm (range: -200 to +130 µm; median: -10 µm; n = 154).

In the meantime, there have been quite a number of publications by several authors comparing the IOLMaster's performance to classical biometry and keratometry or keratoscopy.[22-27] All of these papers confirm the high measurement quality of laser interference biometry.

PCI-Based IOL Calculation

Among echographers, it is well accepted that immersion biometry is more accurate and more reliable than ultrasound biometry performed in contact mode. Contact ultrasound, however, is more popular and more widely used because it is a quick procedure that can easily be applied. Manufacturers of IOLs are of course aware of this fact and determine the A- or ACD constants of their lenses in such a way that optimum results are achieved with contact echometry. Yet an A-constant working well for an AL in contact coupling cannot work if the same eye is measured in immersion. In this case, AL would be longer by some 0.1 to 0.3 mm. A longer eye, however, calls for a weaker IOL power, of the order of 0.3 to 0.9 D. IOL constants for immersion ultrasound, therefore, have to be stronger than the usually given values by up to ≈ 1D. Since the IOLMaster, as has been explained above, emulates an ultrasound immersion measurement, the reasoning for immersion biometry also applies to optical biometry, ie, new (stronger) constants have to be used for the IOLMaster.[8] This is exemplified in Table 14-2, where optimized A-constants of two IOLs (Alcon MA60BM, n = 57; Pharmacia 911A [Pharmacia Corp, Peapack, NJ], n = 116) for the SRK II and SRK/T formulas are listed, separately for ultrasound and the IOLMaster. The optical constants are some 1 D stronger than those for ultrasound. In addition, it can be seen from Table 14-2 that SRK II and SRK/T require different (optimized) A-constants, with differences amounting to 0.5 to 0.7 D.

The necessity of having to calculate new lens constants for optical biometry applies to all IOL power formulas, since it is caused by ALs now measured longer than before. Figure 14-7 shows the necessary changes in the manufacturer's A-constant (118.3) of the Pharmacia lens 911A and five popular IOL formulas in order for them to produce optimum results with ultrasound and IOLMaster ALs. For this lens, the manufacturer's A-constant was nearly optimum if it was used with the IOLMaster and the Haigis formula or with ultrasound in combination with the other formulas. Conversely, the manufacturer's A-constant was too strong for the combination Haigis/ultrasound and too weak for the other formulas in combination with the IOLMaster. This behavior is essentially due to the fact that the Haigis formula emerged from an immersion ultrasound setup, whereas the other formulas originate from an ultrasound world dominated by the contact technique.

> The necessity of having to calculate new lens constants for optical biometry applies to all IOL power formulas, since it is caused by ALs now measured longer than before.

For the derivation of optimized lens constants in the examples above, an outcome analysis was necessary. Using preoperative biometry and keratometry data as well

TABLE 14-2.
Optimized A-Constants

	Alcon MA60BM		Pharmacia 911A	
Manufacturer's A-constant	118.9		118.3	
Formula	Ultrasound	IOLMaster	Ultrasound	IOLMaster
SRK II	119.93	120.96	118.50	119.51
SRK/T	119.25	120.14	118.16	119.05

Optimized A-constants for two IOLs (Alcon MA60Bm, n = 57; Pharmacia 911A, n = 116) and two IOL power formulas (SRK II and SRK/T), optimized for ultrasound and PCI biometry with the IOLMaster. Original patient data courtesy of J. Strobel, University Eye Hospital, Jena, Germany.[28] Note that IOLMaster constants are some 1 D stronger than constants for ultrasound. Also note that SRK II and SRK/T require *different* (optimized) A-constants.

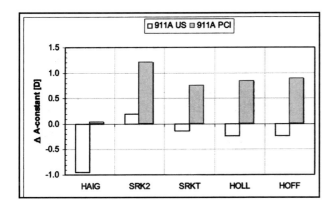

Figure 14-7. Necessary changes in manufacturer's A-constant (118.3) of the Pharmacia lens 911A for ultrasound (US) and IOLMaster (PCI) biometry and 5 IOL power calculation formulas (HAIG: Haigis, SRK2: SRK II, SRKT: SRK/T, HOLL: Holladay [-1], HOFF: HofferQ). Original patient data courtesy of J. Strobel, University Eye Hospital Jena, Germany.[28] For this lens, the manufacturer's A-constant was nearly optimum for the combination IOLMaster/Haigis formula or ultrasound/other formulas. It was too strong for the combination ultrasound/Haigis and too weak for the combination IOLMaster/other formulas.

as stable postoperative refractive results, the very IOL constants were deduced, which produced a mean prediction error of zero for the calculated refraction. We have shown[8] that with proper individualization of lens constants there is virtually no difference between refractive results based on PCI or high precision ultrasound immersion measurements. Similar results have been reported by other groups.[24-27]

User Group for Laser Interference Biometry (ULIB)

At present, optimized IOL constants to be used with the IOLMaster are not readily available from lens manufacturers. Therefore, a user group for optical biometry

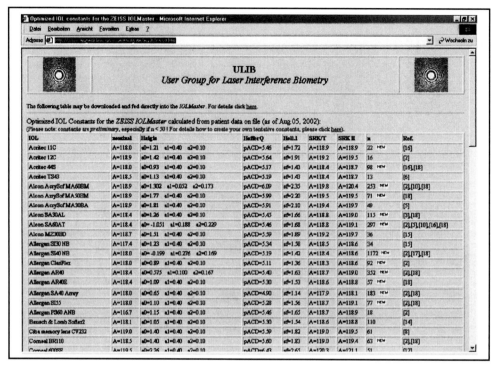

Figure 14-8. Table of optimized IOL constants for the Zeiss IOLMaster published and periodically updated by the User group for Laser Interference Biometry (ULIB, formerly EULIB, www.augenklinik.uni-wuerzburg.de/eulib).

(www.augenklinik.uni-wuerzburg.de/eulib) was founded in 1999 with the intention to collect clinical data and to derive suitable constants for the IOLMaster. The European User group for Laser Interference Biometry (EULIB) soon dropped the word "European" to honor the fact that colleagues from all over the world had joined. Today (as of 2003), the User group for Laser Interference Biometry (ULIB) comprises more than 100 scientists from 19 countries, most of them from the United States, Germany, and Canada. ULIB is intended to serve as a platform to exchange information, news, clinical results, application reports, etc. related to optical biometry and its clinical applications. The main contents on the homepage of the website are:

- a moderated mailing list
- a message board
- news (software notes, bug reports, application notes)
- a download area (form for IOL constants optimization)
- tips and formulas to create individual IOL constants for the IOLMaster
- a continuously updated list of optimized IOL constants for the IOLMaster

This list (Figure 14-8) of optimized IOL constants is one of ULIB's special concerns. As of 2003, it summarizes the results for 38 different IOL types based on 5226 patient datasets, which were provided by 28 surgeons and ophthalmic centers from all over the world and sent to our laboratory for evaluation. A spreadsheet prepared to accept all patient data necessary for constants' optimization is available from the website (www.augenklinik.uni-wuerzburg.de/eulib/dload.htm).

The constants page (www.augenklinik.uni-wuerzburg.de/eulib/const.htm) is periodically updated as we receive new clinical information. The table itself (www.augenklinik.uni-wuerzburg.de/eulib/ioldload.htm) can be downloaded onto a 3.5″ floppy disk and directly fed into the IOLMaster (software version 3.0 or higher required).

These ULIB services have well been accepted by the ophthalmic community. At present, the constants page is accessed more than 800 times per month on average.

Beneficial Use of Optical Biometry

In Myopic Eyes

It has already been mentioned that optical biometry measures along the visual axis since the patient fixates onto the laser spot during measurement. This feature makes PCI biometry superior over echometry in long myopic eyes, where a staphyloma posticum or a deformed back wall is quite common. According to Curtin and Karlin,[29] staphyloma incidence is 1.4% at 27 mm AL, rising to 69.2% at 33 mm. From systematic comparisons of optical and acoustical ALs, it is our clinical impression that these numbers actually are higher. Even normal and short eyes now and then show differences between PCI- and immersion-derived ALs due to shape distortions of the ocular back wall, which can be visualized by an ultrasound B-scan.

Clinical Application

In a myopic eye, an optical AL measurement is more reliable with respect to IOL calculation than an ultrasound measurement. Figure 14-9 gives an example for AL results with both methods: ultrasound measured ≈ 1.6 mm shorter than the IOLMaster. The "true" AL was 28.25 mm, the measurement result obtained by optical biometry. This was confirmed by an ultrasound B-scan and the patient's postoperative manifest refraction. If we had based IOL calculation on the ultrasound value, we would have produced a (myopic) refractive surprise of ≈ 3 to 3.5 D.

In Eyes with Unusual Ocular Media

Another class of applications where PCI-biometry is superior to ultrasound comprises all cases with unusual ocular media, eg, pseudophakic or silicone oil-filled eyes. We have to recall the fact that an ultrasound A-scan does not measure distance, but time-of-flight; likewise, the IOLMaster hardware does not measure distance, but optical path length. In either case, to obtain a distance reading, a propagational speed has to be introduced: velocity of sound for the A-scan instrument, (group) velocity of light for the PCI device. (Actually, the group velocity of light is entered via the group refractive index, which is the ratio between the speeds of light in vacuum and the respective medium.) The differences in sound velocities of different media, eg, materials IOLs are made of, are much more pronounced than the respective differences in the propagational velocities of light. This is to the avail of optical biometry, because corrections that must be applied to such eyes if they are treated as "normal eyes" are much smaller in optical biometry than in ultrasound biometry.

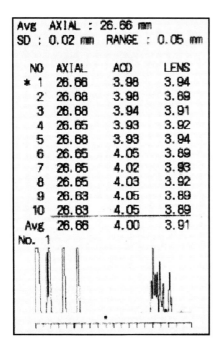

NO	AXIAL	ACD	LENS
* 1	26.66	3.98	3.94
2	26.68	3.98	3.89
3	26.68	3.94	3.91
4	26.65	3.93	3.92
5	26.68	3.93	3.94
6	26.65	4.05	3.89
7	26.65	4.02	3.93
8	26.65	4.03	3.92
9	26.63	4.05	3.89
10	26.63	4.05	3.89
Avg	26.66	4.00	3.91

Avg AXIAL : 26.66 mm
SD : 0.02 mm   RANGE : 0.05 mm
No. 1

Figure 14-9A. Ultrasound and optical biometry in a staphylomatous eye. Ultrasound axial length (Tomey AL-2000 [Tomey Corp., Nagoya, Japan]): AL = 26.66 mm.

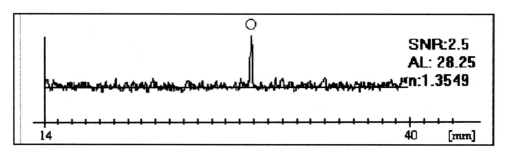

SNR:2.5
AL: 28.25
n:1.3549

14   40   [mm]

Figure 14-9B. PCI axial length (Zeiss IOLMaster): AL = 28.25 mm. The true axial length confirmed by ultrasound B-scan and postop manifest refraction was 28.25 mm. IOL calculation based on the ultrasound value would have led to a refractive surprise of 3.0 to 3.5 D.

Pseudophakic Eyes

If, for example, a pseudophakic eye with a PMMA lens is measured as a normal phakic eye (ie, in "phakic mode" of the A-scan unit), the ultrasound echoes will return earlier than in a truly phakic eye due to the higher sound velocity in the artificial lens. A shorter time-of-flight would make us believe the AL was actually shorter than in reality. Therefore, to make allowance for the higher velocity of sound in the lens, a correction has to be applied to the value obtained in the "phakic mode".

TABLE 14-3.
Correction Factors

Material/Medium	K(PCI) [mm]	K(US) [mm]
Silicone 1 (SLM 1)	+0.12	-1.09
Silicone 2 (MVPS)	+0.12	
Silicone 3 (SLM2)	+0.11	
Memory lens	+0.11	
PMMA	+0.11	+0.35
Acryl (AcrySof)	+0.10	+0.16
Silicone oil	-0.75	-8.79
Phakic	0	0
Aphakic	+0.21	0

Correction factors K in the relation $AL_{true} = AL_{phakic} + K$ between true axial lengths AL_{true} and axial lengths measured in the "phakic mode" AL_{phakic} of the IOLMaster or an ultrasound biometry instrument. Optical correction factors are taken from Haigis.[10] Ultrasound correction factors were deduced in analogy to the User's Manual from the same "average biometric" eye of 23.48 mm used there. Ultrasound velocities were taken from Holladay.[30] The optical correction factors shown here are wired into the IOLMaster.

Clinical Application

If an AL result AL_{phakic} is obtained in the "phakic mode" of an ultrasound biometry instrument (or the IOLMaster), the true axial length (AL_{true}) can be derived from the relation $AL_{true} = AL_{phakic} + K$ with a suitable correction factor K. While this factor for a PMMA lens is typically +0.35 mm for ultrasound (Table 14-3), the PCI value is only +0.11 mm. For a silicone lens, the ultrasound correction is -1.09 mm, opposed to +0.12 mm for the IOLMaster.

Compared to ultrasound, these factors are nearly one order of magnitude smaller for optical biometry, as can be seen from Table 14-3.[10] Also, different IOL materials have optical correction factors very close to each other, contrary to their acoustical counterparts. For the IOLMaster, it is therefore not critical which type of IOL is present in an optical AL measurement. An error caused by choosing the wrong IOL material would be of the order of up to 20 μm, while the same error in ultrasound biometry could cause a difference of up to 1.44 mm and hence a 4 D disaster.

Eyes with Silicone-Filled Vitreous

Not only myopic and pseudophakic eyes benefit from optical biometry, this is also true for silicone oil-filled eyes. Again, as can be seen from Table 14-3, the necessary corrections are about one order of magnitude smaller in optical biometry. We found correlations as high as 98% (unpublished data) between ALs before and with silicone oil endotamponades.

TABLE 14-4. Problem Cases for PCI Biometry	
Anterior Segment	Nystagmus, lid abnormalities, severe tear film problems, keratopathy, corneal scars, mature cataract, PCO
Posterior Segment	Vitreous hemorrhage, membrane formation, maculopathy, retinal detachment
General Conditions	Poor patient cooperation, fixation problems, tremor, respiratory distress

IN CHILDREN

Another advantageous use of optical biometry is the measurement in children. The youngest child whose AL we could measure with the IOLMaster was 1 ½ years old. Quite a number of children could be saved from general anesthesia, which would have otherwise been necessary in order to obtain reliable AL readings. Problems we encountered in this respect were more psychological in nature than technical: how to attract the child's attention to look into the instrument, how to measure the second eye when the first eye was already done, etc.

Optical Versus Acoustical Biometry

Having described the advantages of optical biometry in children and in myopic, pseudophakic, and silicone oil-filled eyes, we have to point out that ultrasound biometry cannot be replaced totally by PCI biometry. Being an optical measurement, light has to get into the eye, reach the retina, and return again to the detector. Any obstruction in the pathway, eg, corneal scars, mature cataracts, dense vitreous hemorrhages, membranes, etc. will render a measurement impossible. But not only a minimum transparency is necessary for laser biometry to work—a certain amount of cooperation on the patient's side is required, too. The examinee has to fixate and be able to keep fixation during 0.4 sec of measurement. Therefore, a maculopathy, respiratory distress, nystagmus, or tremor poses severe problems for optical biometry. Table 14-4 gives an overview of conditions in which PCI biometry might fail.

Any obstruction in the pathway, eg, corneal scars, mature cataracts, dense vitreous hemorrhages, membranes, etc. will render a measurement impossible.

An important question is how many patients actually cannot be measured with the IOLMaster. In one of our first studies,[11,12] we found a percentage of 9% (58 eyes out of 678) of unmeasurable eyes, in another one 12%[8] (16 out of 134 eyes). Other studies report results of 11.2%,[31] 11%,[32] 8%,[26] and 10%.[33] By thoroughly comparing all individual optical scans within a measurement series in order to locate recurrent signals even of low signal-to-noise-ratio (SNR), it is sometimes possible to obtain meaningful results in patients who were not measurable in the first place. Also, there are some procedural tricks (perform optical biometry prior to any other ocular examination, having highly ametropic patients wear their glasses during measurement, deliberately readjust the laser beam out of its center position in order to pass

TABLE 14-5.
Comparison of Ultrasound and Laser Interference Biometry

Ultrasound Biometry	PCI Biometry
Contact via shell (immersion) or xducer (contact coupling)	Non-contact measurement
Infection transfer possible	No infection transfer
Topical anesthesia necessary	No local anesthesia necessary
Measurement more time-consuming	Measurement in 0.4 sec.
Less user friendly	User friendly
Less patient friendly	Patient friendly
Segmental measurements possible	No segmental measurements possible with currently available hardware
All eye segments displayed	Only back wall signal displayed
Adjustment is critical	Defocus uncritical
Contact biometry is less precise than immersion biometry	Measurement quality is comparable to high precision immersion ultrasound
No patient cooperation necessary	Minimum transparency and cooperation necessary

through a non-obstructed area of the cataractous lens, etc.) that help obtain readings from unmeasurable patients. From our experience, some 25 to 30% of these cases can thus be made measurable. In all other cases, ultrasound biometry remains the method of choice.

Furthermore, there is more to biometry than just measuring AL for the purpose of IOL calculation: precise segmental measurements, off-axis and extraocular measurements, measuring thickness of muscles or the optic nerve or of tumors, distance measurements for the location of intraocular foreign bodies, etc. For all of these cases, too, ultrasound biometry has to be applied.

Nevertheless, the determination of AL for refractive purposes is one of the most important biometrical applications, and the advent of laser interference biometry has made this task easier and more comfortable for both patient and examiner. As a non-contact method, it bears no risk of infection transfer and does not need topical anesthesia. Among the drawbacks is the fact that with presently available instrumentation, no segmental measurements are possible in optical biometry. Only the back wall signals are visualized and—unlike in echography—no neighboring ocular structures are displayed. Consequently, very little diagnostic information can be gained from an optical A-scan, contrary to an ultrasonic A-scan. Table 14-5 shows a schematic comparison of advantages and disadvantages of both methods.

Summary

Optical biometry based on partial coherence interferometry has established itself as an alternative to ultrasound biometry as far as AL measurements are concerned. The accuracy of PCI biometry is as high as precision immersion ultrasound and significantly better than standard contact echometry. Laser interference biometry offers definite advantages in children and in myopic, pseudophakic, and silicone-filled eyes. A low percentage of patients, some 5 to 10%, will however remain unmeasureable for optical biometry, and ultrasound biometry will continue to take over its part in these patients as well as in other biometrical applications.

Acknowledgments

The author wishes to thank J. Strobel, head of the University Eye Clinic in Jena, Germany, for providing patient data regarding the Alcon MA60BM and Pharmacia 911A IOLs.

References

1. Mundt GH, Hughes GF. Ultrasonics in ocular diagnosis. *Am J Ophthalmol*. 1956;41:488-498.
2. Hitzenberger CK. Optical measurement of the axial eye length by laser Doppler interferometry. *Invest Ophthalmol Vis Sci*. 1991; 2:616-624.
3. Hitzenberger CK, Drexler W, Dolezal C, Skorpik F, Juchem M, Fercher AF, Gnad HD. Measurement of the axial length of cataract eyes by laser Doppler interferometry. *Invest Ophthalmol Vis Sci*. 1993;34:1886-1893.
4. Findl O, Drexler W, Menapace R, Bittermann S, Fercher AF. Hochpräzisionsbiometrie pseudophaker Augen mittels Teilkohärenzlaserinterferometrie. In: Ohrloff C, Kohnen T, Duncker G, eds. *11. Kongress der DGII 1997*. Berlin: Springer-Verlag; 1998: 120-124.
5. Fercher AF, Roth E. *Ophthalmic laser interferometer. Proceedings of SPIE*. Bellingham, WA: SPIE; 1986: 48-51.
6. Haigis W, Lege BAM. Optical and acoustical biometry. Presented at: ASCRS/ASOA meeting; April 10-14, 1999; Seattle, WA.
7. Haigis W, Lege BAM. Ultraschallbiometrie und optische Biometrie. In: Kohnen T, Ohrloff C, Wenzel M, eds. *13. Kongress d. Deutschspr. Ges. f. Intraokularlinsen-Implant. und refraktive Chirurgie*. Frankfurt: Biermann-Verlag Köln; 2000: 180-186.
8. Haigis W, Lege B, Miller N, Schneider B. Comparison of immersion ultrasound biometry and partial coherence interferometry for IOL calculation according to Haigis. *Graefe's Arch Clin Exp Ophthalmol*. 2000;238:765-773.
9. Haigis W. Optical coherence biometry. In: Kohnen T, ed. *Modern Cataract Surgery*. Basel, Karger; 2002:34:119-130.
10. Haigis W. Pseudophakic correction factors for optical biometry. *Graefe's Arch Clin Exp Ophthalmol*. 2001; 239(8):589-598.
11. Lege BAM, Haigis W. Optical biometry: first clinical experiences. Presented at: ASCRS/ASOA meeting; April 10-14, 1999; Seattle, WA.
12. Lege BAM, Haigis W. Erste klinische Erfahrungen mit der optischen Biometrie. In: Kohnen T, Ohrloff C, Wenzel M, eds. *13. Kongress d. Deutschspr. Ges. f. Intra-okularlinsen-Implant. und refraktive Chirurgie*. Frankfurt: Biermann-Verlag Köln; 1999: 175-179.
13. Fercher AF, Mengedoht K, Werner W. Eye length measurement by interferometry with partially coherent light. *Optics Lett*. 1988;13:186.
14. Zeiss Meditec AG. *IOLMaster User's Manual*. Document 000000-1150.839. 2001;52.

15. Drexler W, Baumgartner A, Findl O, Hitzenberger CK, Sattmann H, Fercher AF. Submicrometer precision biometry of the anterior segment of the human eye. *Invest Ophthalmol Vis Sci.* 1997;38:1304-1313.
16. Findl O, Drexler W, Menapace R, Hitzenberger CK, Fercher AF. High precision biometry of pseudophakic eyes using partial coherence interferometry. *J Cataract Refract Surg.* 1998;24:1087-1093.
17. Pancharatnam S. *Partial polarisation, partial coherence and their spectral description for polychromatic light – part II. Proceedings of the Indian Academy of Sciences.* 1963;57:231.
18. Haigis W, Lege B. First experiences with a new optical biometry device. Presented at: the XVII Congress of the European Society of Cataract and Refractive Surgeons; September, 1999; Vienna.
19. Haigis W, Lege BAM. Akustische und optische Biometrie im klinischen Einsatz. In: Wenzel M, Kohnen T, Blumer B, eds. *14. Kongress der Deutschsprachigen Gesellschaft für Intraokularlinsen-Implantation und refraktive Chirurgie.* Luzern, Schweiz: Biermann-Verlag Köln; 2000; 73-78.
20. Vogel A, Dick HB, Krummenauer F, Pfeiffer N. Reproduzierbarkeit der Messergebnisse bei der optischen Biometrie: Intra- und Interuntersucher-Variabilität. In: Wenzel M, Kohnen T, Blumer B, eds. *14. Kongress der Deutschsprachigen Gesellschaft für Intraokularlinsen-Implantation und refraktive Chirurgie.* Luzern/Schweiz: Biermann-Verlag Köln; 2000; 85-91.
21. Vogel A, Dick HB, Krummenauer F. Reproducibility of optical biometry using partial coherence interferometry: intraobserver and interobserver reliability. *J Cataract Refract Surg.* 2001;27(12):1961-1968.
22. Lam AKC, Chan R, Pang PCK. The repeatability and accuracy of axial length and anterior chamber depth measurements from the IOLMaster. *Ophthal Physiol Opt.* 2001;21(6): 447-483.
23. Santodomingo-Rubido J, Mallen EAH, Gilmartin B, Wolffson JS. A new non-contact optical device for ocular biometry. *Br J Ophthalmol.* 2002;86:458-462.
24. Packer M, Fine IH, Hoffman RS, Coffmann PG, Brown LK. Immersion A-scan compared with partial coherence interferometry: outcomes analysis. *J Cataract Refract Surg.* 2002;28(2):239-242.
25. Connors R, Boseman P, Olson RJ. Accuracy and reproducibility of biometry using partial coherence interferometry. *J Cataract Refract Surg.* 2002;28(2):235-238.
26. Rajan MS, Keilhorn I, Bell JA. Partial coherence laser interferometry vs conventional ultrasound biometry in intraocular lens power calculations. *Eye.* 2002;16:552-556.
27. Kiss B, Findl O, Menapace R, Wirtitsch M, Petternel V, Drexler W, Rainer G, Georgopoulos M, Hitzenberger C, Fercher AF. Refractive outcome of cataract surgery using partial coherence interferometry and ultrasound biometry. *J Cataract Refract Surg.* 2002;28:230-234.
28. Haigis W, Strobel J. Influence of measurement conditions on IOL Constants. Presented at: Symposium on Cataract, IOL and Refractive Surgery of the American Society of Cataract and Refractive Surgery (ASCRS); April 28-May 2, 2001; San Diego, CA. Abstracts, p. 175, 2001.
29. Curtin BJ, Karlin DB. Axial length measurements and fundus changes of the myopic eye. *Am J Ophth.* 1971;71(1):42-53.
30. Holladay JT. Standardizing constants for ultrasonic biometry, keratometry, and intraocular lens power calculation. *J Cataract Refract Surg.* 1997;23:1356-1370.
31. Schrecker J, Strobel J. Optische Achsenlängenmessung mittels Zweistrahl-interferometrie. In: Kohnen T, Ohrloff C, Wenzel M, eds. *13. Kongress d. Deutschspr. Ges. f. Intra-okularlinsen-Implant. und refraktive Chirurgie, Frankfurt 1999.* Biermann-Verlag Köln; 2000; 169-174.
32. Kiss B, Findl O, Menapace R, Wirtitsch M, Drexler W, Hitzenberger C, Fercher AF. Biometry of cataractous eyes using partial coherence interferometry. *J Cataract Refract Surg.* 2002;28:224-229.
33. Verhulst E, Vrijghem JC. Accuracy of intraocular lens power calculations using the Zeiss IOLMaster. A prospective study. *Bull Soc Belge Ophthalmol.* 2001;281:61-65.

15

B-Mode Guided Biometry

Olivier Bergès, MD
Kamal Siahmed, MD
Michel Puech, MD
Francois Perrenoud, MD

Numerous articles dealing with the improvement of intraocular lens (IOL) power calculation have analyzed[1] the efficacy of available formulas, but few have described and compared the different methods to improve the accuracy of keratometry or axial length (AL) measurement.[1-3]

A-mode ultrasonography is widely recognized as the gold standard method for AL measurement (and therefore for IOL power calculation).[1,3-7] The probe is subjectively aligned with the visual axis and the measurement is considered correct when the greatest AL value and the highest amplitude of ultrasound peaks are obtained, indicating perpendicularity to the successive intraocular interfaces of interest. Naturally, precise AL measurement is strictly correlated to the exact recognition of these peaks representing the front surface of the cornea, anterior and posterior lens capsules, and macular vitreoretinal interface. However, correct identification of these landmarks can be difficult in some clinical situations[5] and this technique has some pitfalls:

1. Most ultrasonographers use a contact A-mode technique and it is difficult to avoid a certain amount of pressure on the cornea with indentation by the probe. In most spherical eyes, the anterior chamber depth (ACD) is frequently 0.2 to 0.4 mm less than the measurement obtained by B-mode. This will result in a shorter AL, increased IOL power, and resultant postoperative myopia. To avoid corneal compression, some authors and some manufacturers advise the use of soft tip probes or an immersion technique for AL measurement.

2. In nonspherical (especially myopic) eyes, the presence of a posterior pole staphyloma is the most frequent condition in which a precise AL measurement may not be obtained. The measurement rule that requires obtaining the longest AL values and the highest peaks is no longer valid, and it may be difficult to find three identical measurements for AL (with variations of less than 0.2 mm). An oblique rather than orthogonal interception of the ultrasound beam by the vitreoretinal interface results from the spatial orientation of the posterior pole surface in presence of a myopic staphyloma.[8,9] This may cause a saw-toothed aspect of the peak of the vitreoretinal interface, which prevents precise localization of the foveolar area, and it becomes extremely difficult to select the right echogram (supposed to coincide with the visual axis). In addition, the obliquity of the macular plane to the visual axis in high myopia is responsible for significant AL differences within a small area.

3. The presence of posterior segment pathology (asteroid hyalosis, hyalitis, vitreous hemorrhage, retinal detachment, maculopathy) may also make it more difficult to recognize the correct vitreomacular interface.

Therefore, a new, easy-to-teach method for echographic AL measurement was developed to resolve these problems. The aim of this chapter is to present this new method for AL measurement and to give some results regarding its efficacy. B-mode guided biometry is a more efficient technique to determine the AL for IOL power calculations. This technique also has the added advantages of avoiding corneal indentation by the ultrasound probe and providing two-dimensional images for a better alignment of the ultrasound beam with the visual axis.

B-Mode Guided Biometry Technique

A simplified immersion bath is created using the manually opened eyelid fissure (Figure 15-1) filled with a carbomer 0.2 % ophthalmic gel (Lacrigel [Laboratoires Europhta, Monaco]). The probe's tip is held in suspension within the gel layer, without touching the corneal surface. An optimal axial section of the eye is obtained at a reduced gain (corresponding roughly to tissue sensitivity in standardized echography) and a control vector, seen on the screen as a superimposed dotted line, is aligned with the visual axis on the frozen image. An A-scan biometry consisting of four (or five) highlighted spots representing intraocular interfaces (corneal epithelium, corneal endothelium [the corneal endothelium may not be taken into account], anterior and posterior capsules of the lens and the vitreoretinal interface at the macular region) is reconstructed along the control vector line (Figures 15-2, 15-3, and 15-4). These interfaces (spikes) determine three or four segments (corneal thickness, anterior chamber depth [ACD], lens thickness, vitreal length), and the sound speed (celerity) is different in these different segments. Adequacy of the measurement is ascertained by:

1. Visualization of the anterior and posterior corneal interfaces.
2. Visualization of the anterior lens surface through the iris plane despite gain attenuation, ascertaining that the control vector is within the pupillary area.
3. Alignment of the posterior lens surface with the corneal and anterior lens landmarks.
4. Visualization of the widest posterior lens surface.
5. Visualization of the vitreoretinal interface in a location temporal to the optic nerve head canal.

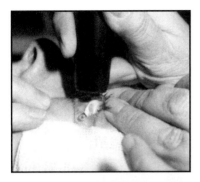

Figure 15-1. B-mode guided biometry technique. Please note the great amount of gel covering the cornea. This simplified immersion technique prevents the probe's tip from touching the cornea.

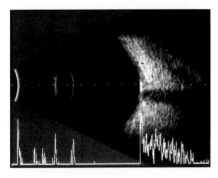

Figure 15-2A. B-mode guided biometry of a normal sized, emmetropic eye. On the B-mode axial section, one can verify the criteria for adequacy of the measurements:
•Visualization of dual anterior and posterior corneal interfaces.
•Visualization of the anterior lens surface through iris plane at reduced gain.
•Visualization of the widest posterior lens surface.
•Visualization of the vitreoretinal interface in a location temporal to the optic nerve head canal (foveola).
•Alignment of the central portions of these structures along a line in coincidence with the visual axis.

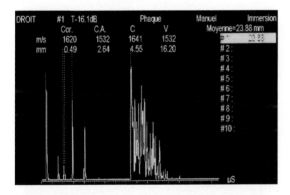

Figure 15-2B. C-vector A-mode reconstructed along the visual axis of the eye. COR = cornea, DROIT = right, Phaque = phakic, Manuel = manual, CA = anterior chamber, C = lens, V = vitreous, Moyenne = mean value.

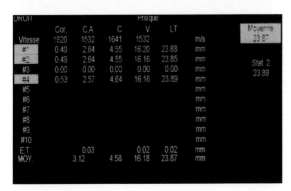

Figure 15-2C. Table of the biometric results. DROIT = right, Phaque = phakic, Cor = cornea, CA = anterior chamber, C = lens, V = vitreous, Vitesse = celerity of the ultrasound beam, Moyenne or MOY = mean value, ET = standard deviation.

Figure 15-3A. B-mode guided biometry of a myopic eye; B-mode axial section. The eyeball is enlarged, ovoid, with an apex of the staphyloma situated on the nasal part of the papilla. The macular plane is oblique to the visual axis, which gives a saw-tooth appearance of the vitreal retinal interface in A-mode. This eyeball morphology is more frequently encountered in smaller eyes with an axial length of 27 mm ± 2 mm.

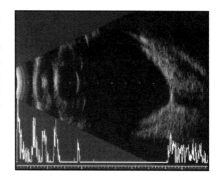

Figure 15-3B. C-vector A-mode reconstructed along the visual axis of the eye (second measurement of three). DROIT = right, Phaque = phakic, Manuel = manual, Cor = cornea, CA = anterior chamber, C = lens, V = vitreous, Moyenne = mean value. Note that even with this eye, which is most difficult to measure due to its shape, the reproducibility of the measurements is excellent.

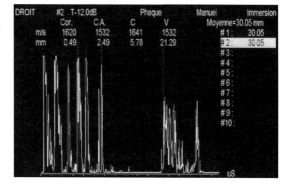

Figure 15-4A. B-mode guided biometry of a very severe myopic eye; B-mode axial section. The staphyloma includes the whole retro-equatorial part of the eyeball with a macular plane quite perpendicular to the visual axis. Irregular appearance of the macula is related to severe choroidal atrophy.

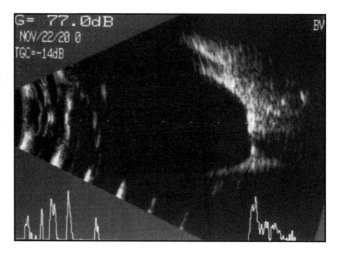

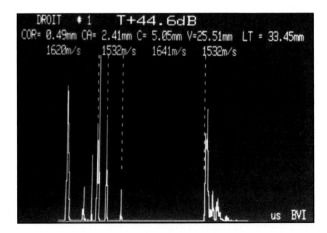

Figure 15-4B. C-vector A-mode reconstructed along the visual axis of the eye. DROIT = right, COR = cornea, CA = anterior chamber, C = lens, V = vitreous, LT = total axial length. It is usually easy to accurately measure this type of very severe myopic eye, but B-mode guided biometry helps in obtaining reproducible measurements.

Clinical Application

Case 1: B-mode axial section of this highly myopic eye (Figure 15-3A) demonstrated the presence of a large staphyloma in the posterior pole and an oblique macular plane. B-mode guided biometry (Figure 15-3B) yielded two reproducible AL measurements of 30.05 mm.

Case 2: B-mode axial section of this severely myopic eye (Figure 15-4A) revealed the presence of a staphyloma involving the whole retro-equatorial part of the eyeball. B-mode guided biometry (Figure 15-4B) yielded an AL measurement of 33.45 mm.

The accuracy of such measurements is 0.1 mm. At least three concordant measurements have to be obtained from three different manually obtained B-mode planes in each eye. The standard deviation for each segment measurement has to be low (< 0.2) (see Figure 15-2C).

Efficacy of B-Mode Guided Biometry

The efficacy of this technique was judged via a study analyzing and comparing the reproducibility and accuracy of a frozen B-mode guided approach versus a classic contact A-scan technique in myopic and nonmyopic eyes (Table 15-1). We will herein only present a summary of the results obtained by our study.[2]

The reproducibility, judged on the mean AL variance, is better with the B-mode guided technique than with the A-scan biometry (Table 15-2). The difference is obvious and statistically significant in the case of myopic eyes, but surprisingly enough, this difference also occurs, even if less pronounced, in nonmyopic eyes (which are supposed to be spherical and easy, and easily measurable with the A-mode technique).

Inter-observer reproducibility was also better with B-mode than with A-mode, as well as the reproducibility of measurements obtained in the same patients during control examinations performed at a variable distance of time. At present, B-mode guided biometry is being used systematically in our departments.

TABLE 15-1.
Repartition of the Two Groups (Myopic and Nonmyopic) of Patients Included in the Study

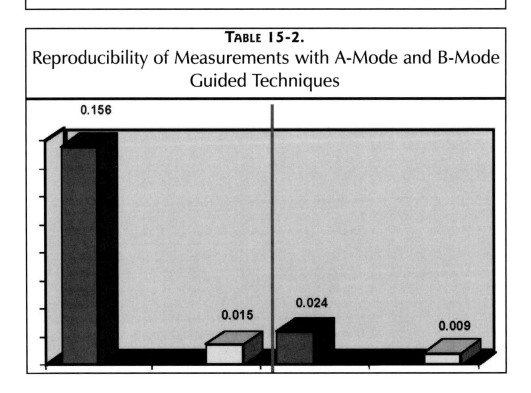

41 non myopic eyes from 21.0 to 23.9 mm
mean value = 22.73
median value = 22.80

42 myopic eyes from 24.3 to 39.8 mm
mean value = 27.32
median value = 26.40

TABLE 15-2.
Reproducibility of Measurements with A-Mode and B-Mode Guided Techniques

0.156

0.015

0.024

0.009

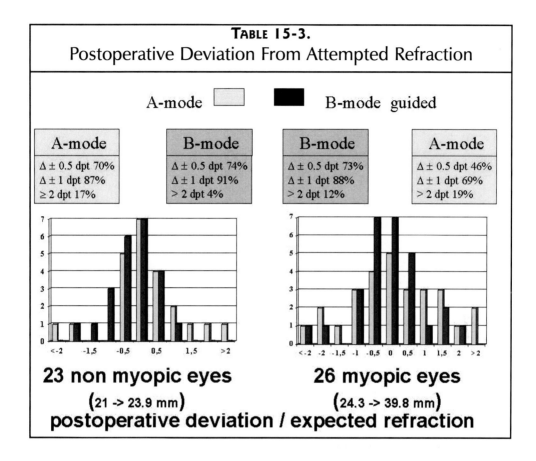

TABLE 15-3.
Postoperative Deviation From Attempted Refraction

23 non myopic eyes
(21 -> 23.9 mm)
26 myopic eyes
(24.3 -> 39.8 mm)
postoperative deviation / expected refraction

The effect of AL accuracy on IOL power calculations was judged by comparing the final postoperative refraction to the expected refraction. Measurements obtained by the B-mode guided technique and by A-scan biometry were compared in the myopic and nonmyopic eyes (Table 15-3). A significantly larger number of eyes measured by B-mode guided biometry had a final refraction within ± 0.50 D from the attempted refraction than the eyes measured by A-mode biometry (74% versus 70% in the nonmyopic group and 73% versus 46% in the myopic group). In addition, deviation greater or equal to 2 diopters (D) was only observed in three cases in the B-mode group for the myopic or nonmyopic eyes, and in nine cases in the A-mode group.

Advantages of B-Mode Guided Biometry

The two main advantages of B-mode guided biometry over A-mode based biometry are:
1. The possibility to identify the visual axis on an axial two-dimensional ultrasonic section of the eye.
2. The prevention of any corneal indentation induced by the ultrasonic probe due to the simplified immersion technique to allow preservation of the actual ACD.

The B-mode also improves identification of the peak representing the anterior lens capsule and differentiates it from the one generated by the iris, a distinction which may sometimes be difficult in A-mode when the ultrasonic beam does not pass right through the center of the pupil.

The distinction between the posterior lens capsule and echogenic interfaces or artifacts within the lens is also more obvious with the B-mode than with the A-mode (Figure 15-5A through C). This allows a more accurate evaluation of the real lens thickness and of the real total AL. This is also the case when distinguishing between the posterior lens capsule and anterior vitreal structures such as floaters or detachment of the anterior hyaloid face.

Clinical Application

B-mode axial section of an eye harboring a very thick cataractous lens (Figure 15-5A) revealed multiple reflections within and behind the lens. B-mode guided biometry allows direct visualization of the posterior lens gate. A correct AL measurement of 22.38 mm was obtained when the gate was positioned over the posterior capsule (Figure 15-5C). The AL measured too short (22.12 mm) when the gate was positioned over some hyperreflective cortex within the lens (Figure 15-5B), and too long (22.45 mm) when the gate was positioned over some retrolental artifacts (Figure 15-5D).

In addition, B-mode biometry offers a good evaluation of the vitreous cavity and of the retina, which is especially important in the presence of a dense cataract and where the fundus cannot be evaluated. It clearly depicts the different stages of age-related macular degeneration (ARMD), the presence of a retinal detachment, or other vitreous opacities such as asteroid hyalosis (Figure 15-6).

Clinical Application

B-mode axial section of an eye with asteroid hyalosis revealed multiple high-reflective echospikes within the vitreous cavity (Figure 15-6A). B-mode guided biometry allows direct visualization and recognition of the retinal spike with three highly reproducible AL measurements of 25.03 mm.

Besides allowing the identification of the correct peaks for AL measurement for IOL power calculation, it gives useful information regarding the potential postoperative visual recovery. In the presence of a choroidal melanoma in the posterior pole, B-mode guided biometry is very effective in measuring the correct AL before starting a conservative treatment such as proton beam therapy (Figure 15-7).

Clinical Application

B-mode axial section of an eye harboring a small melanoma in the macular area (Figure 15-7A). B-mode guided biometry allows the retinal gate to be placed behind the elevated retinal spike with a correct AL measurement of 23.59 mm (Figure 15-7B).

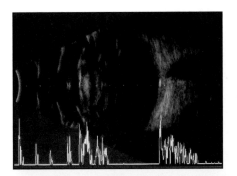

Figure 15-5A. B-mode guided biometry of an eye harboring a very thick and hyperreflective lens. B-mode axial section.

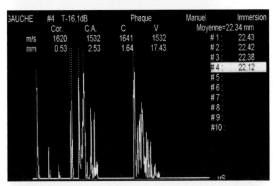

Figure 15-5B. C-vector A-modes reconstructed along the visual axis of the eye, with the highlighted spot corresponding to the posterior capsule being placed on the hyperreflective anterior cortex (which is the more reflective spike). GAUCHE = left, Phaque = phakic, Manuel = manual, Cor = cornea, CA = anterior chamber, C = lens, V = vitreous, Moyenne = mean value.

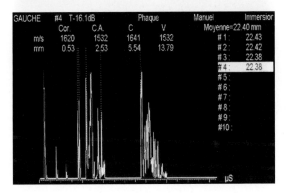

Figure 15-5C. C-vector A-modes reconstructed along the visual axis of the eye, with the highlighted spot corresponding to the posterior capsule being placed on the right posterior capsule.

Figure 15-5D. C-vector A-modes reconstructed along the visual axis of the eye, with the highlighted spot corresponding to the posterior capsule being placed on artifactitious retrolental echoes situated in the vitreous. Note that the lens appears reflective and attenuates the ultrasound beam. Therefore, it is difficult in A-mode to select the right spike corresponding to the posterior capsule of the lens (which is less reflective than the posterior cortex). Note that, due to the high celerity of the ultrasound beam within the lens, a small error in the selection of the lental spikes greatly modifies its thickness (and therefore the total axial length - 22.12 mm to 22.45 mm).

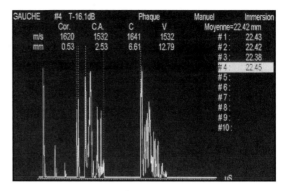

Figure 15-6A. B-mode guided biometry of an eye with very dense asteroid hyalosis. B-mode axial section. Asteroid hyalosis often prevents exact recognition of the vitreoretinal interface with A-mode alone (see the reconstructed A-mode beneath the B-mode picture). This is much easier when one is guided by the B-mode.

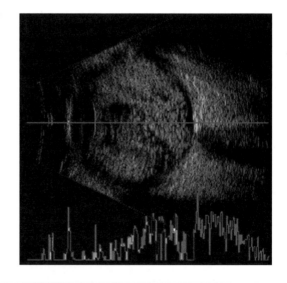

Figure 15-6B. Table of the biometric results. Cor = cornea, CA = anterior chamber, C = lens, V = vitreous, LT = total axial length, Vitesse = celerity of the ultrasound beam, Moyenne or MOY = mean value, ET = standard deviation. Note the extremely good standard deviation (ET) related to total axial length and very low for the measurement of the different intraocular segments obtained after three distinct measurements.

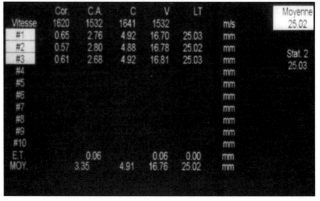

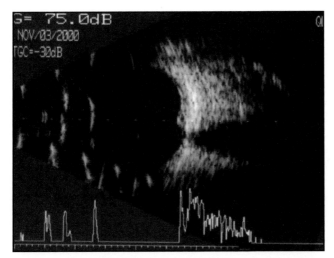

Figure 15-7A. B-mode guided biometry of an eye harboring a small melanoma of the macular region. B-mode axial section. Due to the presence of the tumor in the posterior pole, the B-mode axial section helps to select the correct spike on the reconstructed A-mode echogram. Axial length in this instance is useful to plan conservative treatment and to allow control examinations after treatment.

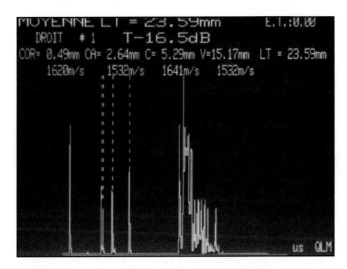

Figure 15-7B. C-vector A-mode reconstructed along the visual axis of the eye. Since flattening of the tumor is expected as a result, axial length measurement must include the retina and the choroid (the tumor) up to the sclera. DROIT = right, Moyenne = mean value, ET = standard deviation, Cor = cornea, CA = anterior chamber, C = lens, V = vitreous, LT = total axial length.

Occasionally, in highly myopic eyes with major deformations of the posterior pole from staphyloma, large variations in the measurement may still be observed, which are mainly due to the oblique orientation of the macular plane to the ultrasonic beam.[9-11]

B-mode guided biometry is highly indicated for:

1. Diagnostic purposes: In the presence of a dense cataract that does not allow sufficient evaluation of the posterior segment.
2. Biometric purposes:
   - Any myopic eye (with an AL exceeding 24 mm with the A-mode technique

- Suspicion of a myopic (or irregular) eyeball with the A-mode technique:
  - Difference of more than 2 mm between the two eyes.
  - Difficulty to find 3 concordant AL measurements within 0.10 mm and a standard deviation of over 0.20 mm.
  - Saw-tooth appearance of the vitreoretinal interface.
  - Difficult to precisely identify the limits of the lens.
  - Presence of real echoes (and not artifacts) in the posterior segment.

The method has very few contraindications (listed below) and would be efficient when partial coherence interferometry may fail. Such conditions may include:

1. Inability to cooperate (fixate), tremor, respiratory distress.
2. Severe tear film problems, keratopathy, corneal scarring.
3. Mature cataract, nystagmus, lid abnormalities.
4. Vitreous hemorrhage, membrane formation, maculopathy, and retinal detachment.

References

1. Olsen T, Nielsen PJ. Immersion versus contact technique in the measurement of axial length by ultrasound. *Acta Ophthalmol.* 1989;67:101-102.
2. Bergès O, Puech M, Assouline M, Letenneur L, Gastellu-Etchegory M. B-mode-guided vector A-mode versus A-mode biometry to determine axial length and intraocular lens power. *J Cataract Refract Surg.* 1998;24:529-535.
3. Giers U, Epple C. Comparison of A-scan device accuracy. *J Cataract Refract Surg.* 1990; 16:235-242.
4. Butcher JM, O'Brien C. The reproducibility of biometry and keratometry measurements. *Eye.* 1991;5:708-711.
5. Hoffer KJ. Preoperative evaluation of the cataractous patient. *Surv Ophthalmol.* 1984;29:55-69.
6. Olsen T, Thim K, Corydon L. Theoretical versus SRK I and SRK II calculation of intraocular lens power. *J Cataract Refract Surg.* 1990;16:217-225.
7. Olsen T. Sources of error in intraocular lens power calculation. *J Cataract Refract Surg.* 1992;18:125-129.
8. Curtin BJ. The natural history of posterior staphyloma development. *Doc Ophthalmol Proc Ser.* 1981;28:207-213.
9. Fernandez-Vigo J, Castro J, Diaz J, Cid MR. Ultrasonic forms of posterior staphyloma. *Ann Ophthalmol.* 1990;22:391-394.
10. Hoffer KJ. Ultrasound velocities for axial eye length measurement. *J Cataract Refract Surg.* 1994;20(5):554-562.
11. Jensen PK, Rask R, Olsen T. Video controlled M-mode biometry. *Acta Ophthalmol Scand.* 1995;73.61-65.

16

Measuring the Ultrasound Axial Length in Biphakic Eyes

Kenneth J. Hoffer, MD

The problem of intraocular lens (IOL) power calculation in eyes that have had corneal refractive surgery is well known. This is due to the inability of our instruments to measure the true corneal power in eyes that have had corneal refractive surgery. Today, with the increasing popularity and proliferation of various designs of phakic refractive IOLs to correct myopia and hyperopia, the question arises whether the phakic IOL will affect the results of normal ultrasound axial length (AL) measurements. The terms *biphakia* and *biphakic* describe the state of a phakic eye containing a refractive IOL regardless of its position in the eye. Since the speed of sound through the different materials of which phakic IOLs are made differs widely and is also different from the average velocity used to measure the eye, a variable but definite error does occur. It is the purpose of this chapter to present one method to correct this error.

Analysis of the Problem

Mathematical analysis of the situation where a phakic eye contains an artificial lens reveals that because the sound traveling through the phakic IOL travels faster or slower than the average speed at which the ultrasound unit is set to measure, the resultant AL provided by the instrument includes a false element in the total measurement. We will call this the phakic IOL thickness error (T_E). To correct the AL, it is obvious that we must subtract this error from the measured AL and then add back the actual true thickness (T_A) of the phakic IOL.

$$AL_{Corrected} = AL_{Measured} + T_A - T_E \qquad (1)$$

We can obtain the true central thickness of the phakic IOL (T_A) from the manufacturer and tables are included in this chapter for phakic IOL thicknesses based on their power for all the commercially available phakic IOLs being implanted today.

We next have to calculate the thickness error (T_E). To do this, we must first start with the fact that there is a directly proportional relationship between the true central thickness (T_A) of the lens and the true sound velocity (V_A) of the material from which the phakic IOL is made and the erroneous thickness (T_E) and incorrect velocity (V_E) used by the ultrasound instrument. This can be expressed as:

$$T_A:V_A::T_E:V_E \text{ which means that:} \qquad (2)$$

$$\frac{T_A}{V_A} = \frac{T_E}{V_E} \qquad (3)$$

If we solve this equation for T_E:

$$T_E = T_A \times [V_E/V_A] \qquad (4)$$

We can now determine the value that must be subtracted by knowing the true sound velocity though the material of which the PRL is made and the velocity of the instrument used for the measurement. Many ultrasound instruments are set at average sound velocities from 1530 meters per second (m/s) to 1560 m/s; often they are set at 1550 m/s. Since 1974, I have used the mathematically determined average velocity of 1555 m/s, published[1] in 1994, which is best for eyes in the normal range of AL. In that paper we also researched the sound velocities through various IOL materials, as shown in Table 16-1.

Now we can replace the erroneous value of thickness in formula 1 with that of formula 4 to produce:

$$AL_{Corrected} = AL_{Measured} + T_A - T_A \times [V_E/V_A] \qquad (5)$$

by pulling T_A from the last two factors, we get:

$$AL_{Corrected} = AL_{Measured} + T_A \times [1 - V_E / V_A]$$

therefore, what must be added to the measured AL is the value $[1 - V_E / V_A]$ multiplied by the actual IOL thickness (T_A).

Effect of the Phakic IOL

Using the ultrasound velocities of the materials from Table 16-1 and the average velocity of 1555 m/s for V_E, we can produce the values to be used shown in Table 16-2. As can be seen, the correction value for IOLs made of collamer is very small and only amounts to adding to the AL 11% of the ICL thickness. For myopic ICLs

TABLE 16-1.
Ultrasound Velocity for Various IOL Materials

Material	Velocity (m/s)
PMMA*	2660
Silicone	980
Collamer**	1740
Acrylic	2026

m/s = meters per second

*Arnold and Guenther[7] reported sound velocity through PMMA at 2710 m/sec, Folds[8] at 2717 m/sec, Kono[9] at 2750 m/sec and Asay, Lamberson and Guenther[10] at 2756 m/sec. But, the most recent report by Hartmann and Jarzynski[11] reported a velocity of 2690 m/sec for PMMA at room temperature. They report that the velocity decreases by 2.5 m/sec for every degree Centigrade increase in temperature. Since these are the values used by the 1989 *Encyclopedia of Polymer Science*,[12] I have chosen 2690 m/sec at room temperature (25º C), adjusted for body temperature (37º C) to 2660 m/sec.

** Value supplied by STAAR Surgical, Inc.

with very thin centers, this will not have a major effect on AL. On the other hand, PRLs made of silicone need 59% of the thickness subtracted from the AL reading and this can be significant, especially in a thick-centered hyperopic PRL (1.09 D error). PMMA lenses require adding 42% of the IOL thickness and acrylic lenses require adding 23% of the central thickness.

Discussion

The phakic IOL does have a variable effect on the ultrasound measurement of the AL. How much is dependent upon the power of the IOL and the material of which it is made. The most effect will be by a silicone PRL in a very high hyperope and the least effect will be in a collamer ICL in a very high myope. Since the central thickness of the only silicone myopic phakic IOLs available at this time (Table 16-2) is always 0.05 mm, it will be necessary to subtract 0.03 mm from the AL in eyes containing a myopic PRL.

The premise of this formulation is based on using an average sound velocity of 1555 m/s for measuring the AL. There may be concern by those who are unable to change the velocity in their A-scan units. We therefore analyzed the effect there would be if the velocity used were as low as 1530 m/s (average velocity for a 30 mm eye) or as high as 1560 m/s (average velocity for a 20 mm eye). As can be seen in Table 16-3, it is fortunate that the effect is so minimal it need not be a concern. In measuring AL in extremes of AL, it is appropriate to adjust for the change in average sound velocity. This can be done by changing the velocity on the A-scan instrument or by the Holladay CALF method[2] of measuring the eye at 1532 m/s (as if it was all aqueous/vitreous) and adding the CALF factor of 0.32 mm. Either way, the correction for the phakic IOL described here is valid. Central thickness for available dioptric ranges of five different phakic IOLs is shown in Tables 16-4 through 16-8.

TABLE 16-2.
Examples of Effective Errors in IOL Power Calculation Measuring Eyes Containing Different Phakic IOLs

Material	1555 m/s	Min CT	TE	IOL Error*	Max CT	TE	IOL Error*
PMMA	+42%	0.160	+0.066	+0.15 D	0.511	+0.212	+0.69 D
Silicone	-59%	0.050	-0.029	-0.07 D	0.573	-0.336	-1.09 D
Collamer	+11%	0.110	+0.012	+0.03 D	1.026	+0.110	+0.35 D
Acrylic	+23%	0.100	+0.023	+0.05 D	0.198	+0.046	+0.10 D

m/s = meters per second, Min CT = minimum central thickness in millimeters, Max CT = maximum central thickness in millimeters, TE = thickness error in millimeters.

*IOL Error = dioptric IOL power error resulting from phakic IOL measurement error, using 2.25 D/mm in long myopic eyes (thin-centered lenses) and 3.25 D/mm in short hyperopic eyes (thick-centered lenses).

TABLE 16-3.
Percentage of the Central Thickness of a Phakic IOL to be Added to the Measured Axial Length Based on the Average Ultrasound Velocity Used to Measure the Eye

Material	1530 m/s	1555 m/s	1560 m/s	Range (mm)	D @ Max CT*
PMMA	+0.4248	+0.4154	+0.4135	0.0113	+12.0
Silicone	-0.5612	-0.5867	-0.5918	0.0306	+15.0
Collamer	+0.1207	+0.1063	+0.1034	0.0173	+20.0
Acrylic	+0.2448	+0.2325	+0.2300	0.0148	-14.0

m/s = meters per second, D = diopters, Max = Maximum, CT = central thickness, mm = millimeters.
*Using 2.25 D/mm in thick-centered IOLs for short hyperopic eyes.

Obviously, these formulations will be tested for accuracy in the real world by comparing the AL measurements both before and after phakic IOL implantation with various materials and powers. But with the limitation of the errors in ultrasound accuracy, this may prove to be more difficult to accomplish. The Zeiss IOLMaster (Zeiss Humphrey Systems, Dublin, CA) may be a better means of comparing once the manufacturer provides a phakic IOL correction capacity.

After the AL has been corrected by the above formula, it is still important to use the correct formula to calculate the IOL power to be implanted. I continue to recommend the Hoffer Q formula[3] for eyes shorter than 22.0 mm, the Holladay 1 formula[4] (not the Holladay 2 formula[5]) for ALs between 24.5 and 26.0 mm, and the SRK/T formula[6] for those longer than 26.0 mm.

TABLE 16-4.

STAAR Surgical Collamer Posterior Chamber ICL, Myopia Models ICM110 to ICM130 (Diameters: 11.0, 11.5, 12.0, 12.5 and 13.0mm), Hyperopia Models ICH115 to ICH130 (Diameters: 11.5, 12.0, 12.5 and 13.0mm)

Phakic IOL Thickness For Each Power			
Diopter	CT mm	Diopter	CT mm
-3.0	0.20	+10.0	0.563
-3.5	0.18	+10.5	0.583
-4.0	0.17	+11.0	0.604
-4.5	0.15	+11.5	0.624
-5.0	0.13	+12.0	0.645
-5.5	0.12	+12.5	0.666
-6 to -20	0.11	+13.0	0.687
+3.0	0.303	+13.5	0.709
+3.5	0.321	+14.0	0.731
+4.0	0.338	+14.5	0.753
+4.5	0.356	+15.0	0.775
+5.0	0.374	+15.5	0.798
+5.5	0.393	+16.0	0.822
+6.0	0.411	+16.5	0.845
+6.5	0.429	+17.0	0.869
+7.0	0.448	+17.5	0.894
+7.5	0.467	+18.0	0.919
+8.0	0.486	+18.5	0.945
+8.5	0.505	+19.0	0.971
+9.0	0.524	+19.5	0.998
+9.5	0.544	+20.0	1.026

Diopter = phakic IOL power in diopters, CT = phakic IOL central thickness, mm = millimeters.

TABLE 16-5.
Medennium Silicone Posterior Chamber PRL

Phakic IOL Thickness for Each Power

Diopter	CT mm		Diopter	CT mm
-30 to -20	0.050		+9.0	0.368
+3.0	0.185		+9.5	0.384
+3.5	0.200		+10.0	0.400
+4.0	0.214		+10.5	0.416
+4.5	0.229		+11.0	0.433
+5.0	0.244		+11.5	0.450
+5.5	0.259		+12.0	0.467
+6.0	0.274		+12.5	0.484
+6.5	0.288		+13.0	0.501
+7.0	0.305		+13.5	0.519
+7.5	0.320		+14.0	0.537
+8.0	0.336		+14.5	0.555
+8.5	0.352		+15.0	0.573

Diopter = phakic IOL power in diopters, CT = phakic IOL central thickness, mm = millimeters.

TABLE 16-6.
Ophtec PMMA Artisan Iris-Fixated Lens

Phakic IOL Thickness for Each Power

Diopter	CT mm		Diopter	CT mm
-3.0 to -23.0	0.140		+7.0	0.393
+2.0	0.209		+7.5	0.412
+2.5	0.227		+8.0	0.432
+3.0	0.245		+8.5	0.451
+3.5	0.263		+9.0	0.471
+4.0	0.281		+9.5	0.491
+4.5	0.299		+10.0	0.512
+5.0	0.318		+10.5	0.532
+5.5	0.336		+11.0	0.553
+6.0	0.355		+11.5	0.574
+6.5	0.374		+12.0	0.596

Diopter = phakic IOL power in diopters, CT = phakic IOL central thickness, mm = millimeters.

TABLE 16-7.
CIBA Vision Acrylic Vivarte Anterior Chamber Lens

Phakic IOL Thickness for Each Power

Diopter	CT mm	Diopter	CT mm	Diopter	CT mm
-7.0	0.189	-13.5	0.100	-20.0	0.100
-7.5	0.179	-14.0	0.198	-20.5	0.187
-8.0	0.163	-14.5	0.182	-21.0	0.169
-8.5	0.148	-15.0	0.165	-21.5	0.152
-9.0	0.132	-15.5	0.149	-22.0	0.135
-9.5	0.116	-16.0	0.133	-22.5	0.118
-10.0	0.100	-16.5	0.116	-23.0	0.100
-10.5	0.196	-17.0	0.100	-23.5	0.154
-11.0	0.180	-17.5	0.184	-24.0	0.136
-11.5	0.164	-18.0	0.167	-24.5	0.118
-12.0	0.148	-18.5	0.150	-25.0	0.100
-12.5	0.132	-19.0	0.134		
-13.0	0.116	-19.5	0.117		

Diopter = phakic IOL power in diopters, CT = phakic IOL central thickness, mm = millimeters.

Summary

A simple model was developed to allow correction of the ultrasound AL of biphakic eyes containing a phakic IOL. It is based on subtracting the erroneous thickness of the IOL from the measured AL and adding back the true thickness. If the eye is measured at an average velocity of 1555 m/s, the following formula can be used, depending upon the material of which the phakic IOL is made:

$$AL_{corrected} = AL_{1555} + X \times T$$

Where T = central thickness of the phakic IOL and X = +0.42 {0.41 - 0.42} for PMMA, -0.59 {0.56 - 0.59} for silicone, +0.11 {0.10 - 0.12} for collamer and +0.23 {0.23 - 0.24} for acrylic.

TABLE 16-8.

Ophthalmic Innovations International Phakic 6 PMMA Anterior Chamber Lens

Phakic IOL Thickness for Each Power

Diopter	CT mm		Diopter	CT mm
+2.0	0.278		-2.0	0.24
+3.0	0.307		-3.0	0.21
+4.0	0.336		-4.0	0.18
+5.0	0.365		-5.0	0.16
+6.0	0.394		-6.0	0.26
+7.0	0.423		-7.0	0.23
+8.0	0.245		-8.0	0.21
+9.0	0.482		-9.0	0.18
+10.0	0.511		-10.0	0.16
			-11.0	0.25
			-12.0	0.23
			-13.0	0.20
			-14.0	0.18
			-15.0	0.16
			-16.0	0.25
			-17.0	0.23
			-18.0	0.20
			-19.0	0.18
			-20.0	0.15
			-21.0	0.26
			-22.0	0.23
			-23.0	0.21
			-24.0	0.18
			-25.0	0.16

Diopter = phakic IOL power in diopters, CT = phakic IOL central thickness, mm = millimeters

References

1. Hoffer KJ. Ultrasound speeds for axial length measurement. *J Cataract Refract Surg.* 1994;20:554-562.
2. Hoffer KJ. *Modern IOL Power Calculations: Avoiding Error and Planning for Special Circumstances*, Focal Points, Clinical Modules for Ophthalmology. American Academy of Ophthalmology, December 1999; 17(12): 1-9.
3. Hoffer KJ. The Hoffer Q formula: A comparison of theoretic and regression formulas. *J Cataract Refract Surg.* 1993;19:700-712. ERRATA: 1994; 20:677.
4. Holladay JT, Prager TC, Chandler TY, Musgrove KH. A three-part system for refining intraocular lens power calculations. *J Cataract Refract Surg.* 1988;14:17-24.
5. Hoffer KJ. Clinical results using the Holladay 2 intraocular lens power formula. *J Cataract Refract Surg.* 2000;26:1233-7.
6. Retzlaff J, Sanders DR, Kraff MC. Development of the SRK/T intraocular lens implant power calculation formula. *J Cataract Refract Surg.* 1990;16:333-340. ERRATA 1990; 16:528.
7. Arnold ND, Guenther AH. Experimental determination of ultrasonic wave velocities in plastics as functions of temperature. *J Appl Polym Sci.* 1966;10:731-743.
8. Folds DL. Experimental determination of ultrasonic wave velocities in plastics, elastomers, and syntactic foam as a function of temperature. *J Acoust Soc Amer.* 1972:52:426-427.
9. Kono R. The dynamic bulk velocity of polystyrene and polymethylmethacrylate. *J Phys Soc* (Japan). 1960;15:718-725.
10. Asay JR, Lamberson DL, Guenther AH. Pressure and temperature dependence on the acoustic velocities in polymethylmethacrylate. *J Appl Phys.* 1969;40:1768-1783.
11. Hartmann B, Jarzynski J. Ultrasound measurements in polymers. *J Acoust Soc Amer.* 1974;56:1469-1477.
12. *Encyclopedia of Polymer Science & Engineering.* Vol 1. New York: Wiley and Sons; 1989.

17

Measuring the Corneal Power

Corneal power is the second most important factor affecting intraocular lens (IOL) power calculations after the axial length (AL). A keratometric error of 1 diopter (D) affects the postoperative refraction by approximately the same amount.

Manual Keratometry

Manual keratometry is the most commonly used method to measure the corneal curvature.[1] It is fast, easy, cheap, and very accurate in most average cases.

The patient is seated behind the keratometer with the chin well positioned in the chin rest and the forehead in touch with the head rest. The keratometer is directed towards the eye to be examined while the occluding shield is placed in front of the opposite eye. The keratometer is focused on the central portion of the cornea using the focusing knots. At this time, the central ring appears as one circle. The left drum is rotated to superimpose the (+) signs and a horizontal measurement is taken (Figure 17-1). The right drum is then rotated to superimpose the (-) signs and a vertical measurement is taken (Figure 17-2).

IOL power formulas rarely use the horizontal (K_1) and the vertical (K_2) readings. They usually call for the average value (K):

$$K = 0.5 (K_1 + K_2)$$

It is important to remember that the keratometer has to be calibrated every 6 months. A set of calibration balls with a known radius of curvature can be used. The horizontal and vertical drums are then set at the same value.

Figure 17-1. The technician rotates the left drum to obtain a horizontal measurement.

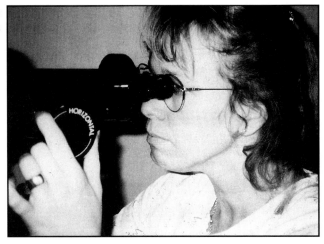

Figure 17-2. The technician rotates the right drum to obtain a vertical measurement.

Automatic Keratometry

An automatic keratometer is designed to give accurate, objective, and reproducible measurements of the K readings.[1] The technique consists of seating the patient behind the automatic keratometer or a keratometer/refractor with the chin resting in the chin cup and the forehead against the forehead rest. The eye to be examined is selected. The patient is instructed to look at the center of the red fixation light and to hold the head steady. Using the control buttons, the light is focused on the central cornea and a measurement is taken. A print-out is obtained for safekeeping in the patient's record. It provides the K readings from the central cornea in the steepest and flattest meridians; the readings are obtained in diopters and in millimeters.

Corneal Topography

A corneal analysis system is used to evaluate central and peripheral corneal curvature.[2,3] The patient is seated behind a corneascope that projects a 16-ring conical placido disc on the cornea. A small camera captures the image reflection and transmits it to a computer that converts the data into a series of color graphic displays.

Although a regular keratometer gives accurate and reproducible measurements in most cases, corneal topography can be helpful in certain cases:
1. When the cornea is flatter than 40 D or steeper than 46 D.
2. When the surgeon wants to better evaluate the pre-existing astigmatism.
3. When the cornea is irregular, ie, after trauma or in the presence of a concomitant keratoconus.
4. When some type of corneal surgery has already been performed, ie, radial keratotomy, laser photorefractive keratectomy, or laser in situ keratomileusis (LASIK).

Modifying the Corneal Power Readings

In certain cases, the measured K readings have to be modified because of concomitant or previous corneal surgery. The modifications are needed to improve IOL power calculations accuracy.

CONCOMITANT ASTIGMATIC KERATOTOMY

Concomitant astigmatic keratotomy and implant surgery are being performed successfully to reduce pre-existing astigmatism. Corneal topography will detect the correct amount of astigmatism and its axis prior to surgery.[4-5]

Astigmatic keratotomy flattens the steep corneal meridian. However, this flattening is associated with a slight to moderate steepening of the cornea 90 degrees away. This is known as the coupling effect. If the amount of flattening is equal to the amount of steepening 90 degrees away, then the average K readings remain the same, and the power of the IOL remains the same. In fact, the amount of flattening slightly exceeds the amount of steepening, ending in a flatter cornea, which affects the IOL power calculations. Ideally, the cornea is remeasured in the operating room after the keratotomy and the IOL power is recalculated. However, if the surgical microscope is not equipped with a keratometer or if the surgeon does not want to redo the calculations in the operating room, then the expected corneal curvature has to be calculated. I suggest subtracting 0.25 D from the average preoperative K readings for every diopter of astigmatism to be corrected.

Clinical Application

Preoperative K readings: 41.00/45.00 D
Average preoperative K: 43.00 D
Plan: Astigmatic keratotomy to correct 4.00 D
Expected postoperative K: $43.00 - (0.25 \times 4) = 42.00$ D

This last value of 42.00 D is used in the IOL power calculations.

Corneal Power Adjustment in the Combined Procedure of Corneal Transplant, ECCE, and Lens Implantation

In the presence of a cataract with marked endothelial dystrophy, causing corneal edema,[6] the surgeon has three choices:

1. Removing the cataract first, followed by a corneal transplant. This procedure can be done if the cornea is still relatively clear. Variations in the final refraction can be encountered, depending on the corneal transplant dioptric power.
2. Performing a corneal transplant first, followed a few months later by cataract and implant surgery. The final refraction is more predictable because the power of the corneal transplant is known at the time of cataract surgery and any induced astigmatism from the corneal transplant can be corrected during the second surgery. However, this scheme requires two surgeries and a long period of visual rehabilitation.
3. With combined corneal transplant, ECCE, and lens implant surgery, the final refraction is not as predictable, but it eliminates second surgery, offers a faster visual recovery, and avoids additional risks to the graft's endothelium.

With this triple procedure, there is a lack of accuracy in the IOL power calculations due to the unknown power of the new cornea.[6-7] The power of a cloudy cornea can be difficult to measure preoperatively. Furthermore, such a measurement can be inaccurate; a scarred cornea will yield weaker K readings. In such cases, both eyes should be measured, and if there is a large discrepancy between the two, it is best to use the K readings of the opposite cornea. Also, the surgeon should remember that a corneal button 0.2 mm larger than the recipient's bed will avoid any major myopic inducement. If the button is 0.2 to 0.5 mm larger than the recipient's bed, the central cornea becomes steeper, causing a myopic shift of 1.00 to 2.00 D.

Corneal Power Adjustment After Corneal Refractive Surgery

This subject is discussed in detail in Chapter 18.

References

1. Shammas HJ. *Atlas of Ophthalmic Ultrasonography and Biometry*. St. Louis, MO: CV Mosby Co.; 1984: 273-308.
2. Rowsey JJ, Reynolds AE, Brown R. Corneal topography: corneascope. *Arch Ophthalmol.* 1981;99:93-100.
3. Koch DD, Haft EA. Introduction to corneal topography. In: Sanders DR, Koch DD, eds. *An Atlas of Corneal Topography*. Thorofare, NJ: SLACK Incorporated; 1992.
4. Axt JC. Longitudinal study of postoperative astigmatism. *J Cataract Refract Surg.* 1987;13:381-388.
5. Stainer GE, Binder PS, Parker WT, Perl T. The natural and modified course of post-cataract astigmatism. *Ophthalmic Surg.* 1982;13:822-827.
6. Lindstrom RL, Harris WS, Doughman DJ. Combined penetrating keratoplasty, extracapsular extraction and posterior chamber lens implantation. *American Intra-Ocular Implant Society Journal.* 1981;7:130-132.
7. Binder PS. Secondary intraocular lens implantation during or after corneal transplantation. *Am J Ophthalmol.* 1985;99:515-520.

Clinical Application, continued

Case 2

A 59-year-old female presented with a cataract in her left eye 2 years after LASIK surgery. The following measurements were obtained from OS:

- Refraction (Rs_{post}): -4.50 - 0.75 x 155° = 20/80
- K readings (K_{post}): 37.75 / 39.00 D [Average K_{post} = 38.37 D]
- Corrected K readings ($K_{c.cd}$): $K_{c.cd}$ = 1.14 K_{post} - 6.8 = 36.94 D
- Axial length: 26.66 mm
- No information from the previous refractive surgery could be obtained.

The modified Hoffer Q formula called for a 20.71 D for emmetropia. Aiming for emmetropia, a 20.5 D AcrySof MA30AC implant was inserted in the capsular bag. Six months after surgery, the refraction was: -0.50 Sph = 20/20.

Case 3

A 54-year-old male presented with a cataract in his right eye 3 years after two LASIK surgeries on the same eye. The following measurements were obtained from OD:

- Refraction (Rs_{post}): -5.75 - 4.75 x 16° = 20/100
- K readings (K_{post}): 38.50 / 40.50 D [Average K_{post} = 39.50 D]
- Corrected K readings ($K_{c.cd}$): $K_{c.cd}$ = 1.14 K_{post} - 6.8 = 38.23 D
- Axial length: 29.09 mm
- No information from the previous refractive surgeries could be obtained.

The modified Hoffer Q formula called for a 12.22 D for emmetropia. Aiming for a myopic refraction of -1.50 to -2.00 Sph, a 15.0 D AcrySof MA60AC implant was inserted in the capsular bag. Two months after surgery, the refraction was: -0.75 -2.00 x 04° = 20/25 (Spherical equivalent of -1.75 Sph).

Using the modified Hoffer Q formula, the final refraction was within ± 0.50 D of the expected result in Case 1 and Case 3, and within ± 1.00 D in all three cases.

References

1. Koch DD, Liu JF, Hyde LL, Rock RL, Emery JM. Refractive complications of cataract surgery after radial keratometry. *Am J Ophthalmol.* 1989;108:676-682.
2. Lyle WA, Jin GJ. Intraocular lens prediction in patients who undergo cataract surgery following previous radial keratometry. *Arch Ophthalmol.* 1997;115:457-461.
3. Seitz B, Langenbucher A, Nguyen NX, Kus MM, Kuchle M. Underestimation of intraocular lens power for cataract surgery after myopic photorefractive keratectomy. *Ophthalmology.* 1999;106:693-702.
4. Gimbel H, Sun R, Kaye GB. Refractive error in cataract surgery after previous refractive surgery. *J Cataract Refract Surg.* 2000;26:142-144.
5. Feiz V, Mannis MJ, Garcia-Ferrer F, Kandavel G, Darlington JK, Kim E, Caspar J, Wang JL, Wang W. Intraocular lens power calculation after laser in situ keratomileusis for myopia and hyperopia: a standardized approach. *Cornea.* 2001;20:792-797.

6. Holladay JT. IOL calculations following radial keratotomy surgery. Questions and answers. *Refractive & Corneal Surg*. 1989;5:36.
7. Seitz B, Langenbucher A. Intraocular lens power calculation in eyes after corneal refractive surgery. *J Refract Surg*. 2000;16:349-361.
8. Shammas HJ. *Intraocular Lens Power Calculations: Avoiding the Errors*. Glendale, CA: The News Circle Publishing House; 1996:117-118.
9. Shammas HJ. Intraocular lens power calculations. In: Azar DT, ed. *Intraocular Lenses in Cataract and Refractive Surgery*. Philadelphia: WB Saunders; 2001: 60-61.
10. Gimbel HV, Sun R. Accuracy and predictability of intraocular lens power calculations after laser in situ keratomileusis. *J Cataract Refract Surg*. 2001;27:571-576.
11. Odenthal MTP, Eggink CA, Melles G, Pameyer JH, Geerards AJM, Beekhuis WH. Clinical and theoretical results of intraocular lens power calculation for cataract after photorefractive keratectomy for myopia. *Arch Ophthalmol*. 2002;120:431-438.
12. Hamed AM, Wang L, Misra M, Koch D. A comparative analysis of five methods of determining corneal refractive power in eyes that have undergone myopic laser in situ keratomileusis. *Ophthalmology*. 2002;109:651-658.
13. Shammas HJ, Shammas MC, Garabet A, Kim JH, Shammas A, Labree L. Correcting the corneal power for IOL power calculations after laser in situ keratomileusis (LASIK). *Am J Ophthalmol*. 2003;136:426-432.
14. Zeh WG, Koch DD. Comparison of contact lens over-refraction and standard keratometry for measuring corneal curvature in eyes with lenticular opacity. *J Cataract Refract Surg*. 1999;25:898-903.
15. Holladay JT, Prager TC, Chandler TY, Musgrove KH. A three-part system for refining intraocular lens power calculations. *J Cataract Refract Surg*. 1988;14:17-24.
16. Hoffer KJ. The Hoffer-Q formula: a comparison of theoretic and regression formulas. *J Cataract Refract Surg*. 1993;19:700-712.
17. Retzlaff J, Sanders DR, Kraff MC. Development of the SRK/T intraocular lens implant power calculation formula. *J Cataract Refract Surg*. 1990;16:333-340.
18. Haigis W. Einflub der Optikform auf die individuelle Anpassung von Linsenkonstanten zur IOL-Berechnung. In: Rochels R, Duncker GIW, Hartmann C, eds. *10. Kongreb der Deutschen Gesellschaft fur Intraokularlinsenimplantation*. Berlin: Springer; 1997;281-287.
19. Olsen T, Thim K, Corydon L. Accuracy of the newer generation intraocular lens power calculation formulas in long and short eyes. *J Cataract Refract Surg*. 1991;17:187-193.

19

Selecting the Proper Intraocular Lens Power

The selection of the proper intraocular lens (IOL) power prior to surgery is an important process and should not be rushed. It is best performed a few days prior to surgery and not at the last minute in the operating room. Careful discussion with the patient will enlighten the surgeon as to the patient's expectations. Some patients know exactly what they want, while others will leave the decision up to their ophthalmologist and it becomes incumbent upon the surgeon to make the proper decision.

Target Postoperative Refraction

Some ophthalmologists advocate emmetropia for all their patients, while others routinely advocate slight myopia.[1-10] Studies have shown that a small amount of myopic astigmatism can enhance the depth of focus of the pseudophakic eye, with adequate 20/30 visual acuity for both near and distance fixation, thus providing spectacle independence. However, not all eyes, or all patients, are the same. The surgeon has to review the patient's needs and expectations, the status of the fellow eye, the IOL power for emmetropia, and then choose the IOL power accordingly.

EMMETROPIA AND AMETROPIA

Emmetropia denotes the absence of any refractive error, while *ametropia* denotes the presence of a refractive error.

Most formulas automatically measure the IOL power needed for emmetropia, which will allow clear distant vision. Also, most calculators and computer programs will give a range of expected postoperative refractive errors with the use of an IOL of a known power; this is helpful if some ametropia is desired. If such a program is not available, the IOL power for ametropia can be calculated by adding the desired refractive error at the cornea to the keratometric readings and using the same formula.

Clinical Application

A patient with an axial length (AL) of 23.5 mm and K readings of 43.5 diopters (D) requires a 20.0 D posterior chamber lens for emmetropia. If a postoperative refraction of -0.50 D is desired, and the K readings are adjusted to 43.50 – 0.50 = 43.0 D, that will require an IOL power of 20.7 D.

Isometropia

Isometropia is equality of refractive error. The isometropic IOL power is calculated with the regular formula used for emmetropia with one modification: the contact lens refractive power of the fellow eye is added to the K readings of the eye to be operated. For example, if the fellow eye has -2.00 D of myopia at the cornea and you decide to make the operated eye, which has average K readings of 44.00 D, isometric with the fellow eye, you enter into the formula the adjusted K readings of 44.00 – 2.00 = 42.00 D. The isometropic IOL power is rarely used because the surgeon has the opportunity to decrease the refractive error during surgery. Patients can easily tolerate a 1.5 to 2.0 D difference in refractive error between the two eyes, with no risks of asthenopia and/or diplopia.

Iseikonia

Iseikonia is the equality of retinal image size.[11-14] To achieve iseikonia, the two eyes have to have the same posterior focal length. In emmetropic eyes undergoing cataract surgery, iseikonia is usually always preserved when IOL calculations aim towards emmetropia or even slight myopia. In eyes with large refractive errors, the problem becomes more complicated and specific formulas are needed for proper calculations (see Chapter 7). The majority of the patients can ignore aniseikonia of up to ± 5%, which reflects a refractive error variation of ± 2.50 D.

The iseikonic IOL power is calculated and seriously considered when the fellow eye is normal with high myopia or high hypermetropia, has good vision, and does not require surgery. The importance of the isometric and iseikonic IOL power calculations decreases tremendously if the fellow eye has a cataract and surgery is contemplated in the near future.

The importance of the isometric and iseikonic IOL power calculations decreases tremendously if the fellow eye has a cataract and surgery is contemplated in the near future.

Patients' Needs and Expectations

Patients' expectations usually coincide with their needs and they should certainly be accommodated accordingly.[15-19] The selection of the IOL power and the expected postoperative refractive error should be discussed with the patient prior to surgery.

IOL power selection in emmetropes is relatively easy and noncontroversial in most cases. Young, active emmetropes want to remain so, while older, sedentary patients might enjoy slight myopia. The surgeon has also to remember that these emmetropes do not tolerate any error in IOL power calculations; they might reluctantly accept induced myopia, but they will rarely accept induced hyperopia.

In an emmetropic eye, the surgeon expects to use an IOL with a power ranging from 19 to 21 D.

Occasionally, IOL power calculations will call for a higher power than expected. Nuclear sclerosis is known to induce a myopic shift, and it is not unusual for a hyperopic eye to become emmetropic with the progression of the cataract. The choice will still be to aim for emmetropia. However, the surgeon should remember the high margin of error in these short eyes.

IOL Selection in the Myopic Eye

Myopes with a refraction ranging between -2.00 and -4.00 Sph usually use their prescription for distance vision and remove their glasses for reading, contrary to higher myopes and hyperopes who are used to wearing their glasses for distance and reading.

Older, sedentary patients usually prefer to remain nearsighted, while young active myopes are more likely to request emmetropia. Surgeons should not aim for "bull's eye" emmetropia, because they can push the patients into hyperopia in case of a minor error in IOL power calculations. Instead they should aim toward a -0.50 to a -1.00 Sph, a refraction that most myopes seem to enjoy the most for distance vision and reading of large print. Different scenarios are herein presented to assist the ophthalmologist in making the correct choice. For the sake of discussion, moderate myopia will range from -2.00 to -4.00 Sph, while high myopia will exceed -4.00 Sph.

BINOCULAR CATARACTS WITH BILATERAL MODERATE MYOPIA

	OD	OS
Vision	20/70	20/80
Refraction	-3.00	-3.00
Cataract	Yes	Yes

Both eyes are myopic and both will need cataract surgery. The choices are:
1. To aim for -2.50 Sph in OU. A -2.50 D of myopia will keep the status quo for this elderly patient. The patient will continue reading without glasses and will use single vision glasses for distance when needed.
2. To aim for -1.00 to -1.50 Sph in the dominant eye and -2.00 to -2.50 Sph in the other eye. This combination will provide functional and reading vision without much anisometropia between the two eyes.
3. To aim for emmetropia in the dominant eye and -2.00 to -2.50 Sph in the other eye. The difference between the two eyes might cause a certain degree of anisometropia.
4. To aim for emmetropia in both eyes.

Choices 1, 2, and 3 might be considered in older, sedentary patients, while choice 4 would be more appropriate for young and active patients.

MONOCULAR CATARACT WITH BILATERAL MODERATE MYOPIA

	OD	OS
Vision	20/20	20/80
Refraction	-2.50	-3.00
Cataract	No	Yes

Both eyes are myopic and only one eye has a cataract. The cataractous eye is scheduled for surgery. The choices in this case are:

1. To aim for -2.00 to -2.50 Sph. This will keep the status quo, eliminating the need for reading glasses. It does not cause any anisometropia or any significant amount of aniseikonia.
2. To aim for -0.75 Sph to -1.00 Sph. The -0.75 D of myopia is a compromise between isometropia and emmetropia. It provides patients with adequate functional vision around the house, but they will still need glasses for distance vision and possibly for reading too.
3. To aim for emmetropia. Emmetropia provides active patients with clear distance vision, but the difference between the two eyes can be intolerable. This choice is viable if the patient insists on emmetropia and if the fellow eye can be fitted with a contact lens or treated with refractive surgery. Occasionally, some patients will accept monocular vision, especially if the cataractous eye is the dominant one. They can then use the operated dominant eye for distance and the nondominant eye for reading.

Choices 1 and 2 are best suited for older, sedentary patients, while choice 3 would be more suited for young, active patients.

MONOCULAR CATARACT IN ONE MODERATELY MYOPIC EYE

	OD	OS
Vision	20/20	20/80
Refraction	Plano	-2.50
Cataract	No	Yes

The right eye is emmetropic with good uncorrected vision, while the left eye is -2.50 myopic with decreased vision due to the presence of a cataract. The cataractous left eye is scheduled for surgery. AL measurement reveals a 1 mm difference between the two eyes. The clinical history might reveal that this patient never wore glasses. This can only mean that the patient always had monocular vision, using the dominant right eye for distance and the nondominant left eye for reading. The choices in this case are:

1. To aim for -2.00 to -2.50 Sph. A myopic refraction will also prevent any large amount of induced aniseikonia. This choice will keep the status quo and a very happy patient.
2. To aim for emmetropia. Although many surgeons will opt first for this choice, it will be a mistake in this case. This choice is, of course, the preferred choice if the myopia in the left eye is secondary to the cataract development, and not to a difference in the AL. This choice will also be preferred if the myopic left eye, with the cataract, is the dominant eye.

BINOCULAR CATARACTS WITH BILATERAL HIGH MYOPIA

	OD	OS
Vision	20/70	20/80
Refraction	-8.00	-8.00
Cataract	Yes	Yes

Patients with high myopia develop cataracts at a younger age than emmetropes and hyperopes. The surgeon often has to deal with young, active patients and their high expectations. The choices in this case are:

1. To aim for plano to -0.50 Sph in the dominant eye and for -1.50 to -2.00 Sph in the fellow eye to free the patient from glasses. However, reaching the planned postoperative refraction might be more difficult in long eyes because of the accentuated curvature of the posterior pole. Furthermore, all available formulas are less accurate in long eyes. In aiming for a -0.50 D of myopia, the surgeon keeps a comfortable margin of error between plano and -1.00 Sph.

2. To aim for emmetropia in both eyes. It is not unusual for such patients to insist on bilateral emmetropia. This choice would be popular with patients who have been wearing corrective contact lenses for distance vision. The surgeon must remember that some young high myopes have not yet experienced presbyopia and might expect good uncorrected vision for distance and near. Furthermore, in aiming for "bull's eye" emmetropia, a slight error in the IOL power calculations could result in some degree of hyperopia and a very unhappy patient. A good discussion with the patient is primordial in this case.

3. To aim for -2.00 to -2.50 Sph in both eyes. This choice is more suitable for older patients that have always been wearing glasses for distance vision. These patients will enjoy reading without any correction and will wear their new, thinner glasses for driving or watching TV.

MONOCULAR CATARACT WITH BILATERAL HIGH MYOPIA

	OD	OS
Vision	20/20	20/80
Refraction	-8.00	-9.00
Cataract	No	Yes

The cataractous left eye is scheduled for surgery. Young patients might want emmetropia in the operated eye. A good discussion in this difficult case is the first priority. The patient can be offered the following choices:

1. To aim for emmetropia or slight myopia in the left eye. It is the dream of most young myopes to have clear, uncorrected vision and this is definitely a golden opportunity for this young executive. Aiming for emmetropia will most certainly create anisometropia and aniseikonia. This can be avoided by decreasing the myopia in the right eye either with a contact lens or with corneal surgery. More daring surgeons will even propose a clear lensectomy to the right eye. However, if the operated eye is the dominant one, patients might be very sat-

isfied with one emmetropic eye and ignore the uncorrected blurry vision from the other eye. This is the preferred choice for young and active patients.
2. To aim for isometropia, leaving the operated eye with 2 D less myopia than the fellow eye. This will avoid a significant degree of anisometropia and aniseikonia. However, the patient is missing the chance to get rid of the glasses. This might be an acceptable choice for older, sedentary patients.

MONOCULAR CATARACT IN UNILATERAL AXIAL MYOPIA

	OD	OS
Vision	20/20	20/80
Refraction	-0.50	-11.00
Cataract	No	Yes

Unilateral axial myopia is a rare congenital anomaly that causes anisometropia and amblyopia in the affected eye. It remains undetected throughout childhood. Patients, usually in their 40s and 50s, present to the ophthalmologist when they lose vision in the affected eye because of the cataract's progression. It is not unusual for the myopia in the nondominant eye to go unnoticed due to the opaque media, and when it is unsuspected, an uninformed technician can disregard the long AL measurement in the affected eye, blame it on a difficult A-scan measurement, and record a reading similar to the normal fellow eye. This will of course result in a very large postoperative refractive error. AL measurements will show a difference of over 3 mm between the two eyes. The choices in this case are as follows:
1. To aim for emmetropia or a slight myopia. Emmetropia or slight myopia will restore isometropia and partial vision; it also increases the field of vision. Emmetropia or slight myopia will not restore iseikonia, but this does not seem to be a problem because of the associated amblyopia.
2. To aim for -2.00 to -2.50 Sph. The patient uses monocular vision. If the left eye recovers adequate vision, it will improve the reading vision without wearing glasses. However, these eyes are usually amblyopic and vision rarely improves better than 20/60.
3. To aim for -7.00 or -8.00 Sph. Leaving the eye highly myopic is not a viable alternative in these young patients.

IOL Power Selection in the Hyperopic Eye

Hyperopic patients look forward to emmetropia after the cataract surgery. Older, sedentary patients might even enjoy a certain degree of myopia. For the sake of discussion, moderate hyperopia will range from +2.00 to +4.00 Sph, while high hyperopia will exceed +4.00 Sph.

BINOCULAR CATARACTS WITH BILATERAL MODERATE HYPEROPIA

	OD	OS
Vision	20/50	20/80
Refraction	+3.00	+3.00
Cataract	Yes	Yes

Both eyes will need cataract surgery. The choices in this case are:
1. To aim for emmetropia in both eyes. The patient will only need glasses for reading.
2. To aim for +0.50 to +1.00 Sph in the left eye. This option is to be considered only if one eye is to be operated, leaving the other eye with 20/40 to 20/60 vision. Leaving the operated eye slightly hyperopic will avoid anisometropia and aniseikonia. Occasionally an older patient is reluctant to have both eyes operated, especially if the anticipation of vision improvement is not certain due to the concomitant presence of macular degeneration.

MONOCULAR CATARACT WITH BILATERAL HIGH HYPEROPIA

	OD	OS
Vision	20/80	20/30
Refraction	+8.00	+8.00
Cataract	Yes	No

This is an unusual case that requires some discussion with the patient as to the best modality of treatment. The choices in this case are as follows:
1. To aim for emmetropia. The patient will certainly enjoy good uncorrected vision, especially if the operated eye is the dominant one. However, emmetropia will cause anisometropia and aniseikonia that have to be addressed and discussed with the patient. To avoid these problems, the left eye can be fitted with a contact lens, if the patient can tolerate it. Another possibility is for the patient to ignore the relatively blurred uncorrected vision in the nonoperated eye and depend on the good uncorrected vision in the operated eye. The nonoperated eye will usually show signs of an early cataract and will have to be operated on sooner or later; the patient might opt for "sooner" rather than "later".
2. To aim for +3.00 Sph to +4.00 Sph. The patient can tolerate 2.00 to 3.00 D of anisometropia without causing any aniseikonia. This solution can be considered if the patient wants to keep wearing the glasses and wants no contact lens or surgery to the other eye.

IOL Power Selection in Presence of Capsular Rupture with or without Vitreous Loss

Most surgeons calculate the IOL power for emmetropia using their favorite foldable implant placed in the capsular bag. If the capsule ruptures, the surgeon might opt to change the surgical plan.

- In case of a capsular rupture without vitreous loss, the surgeon will probably insert the same implant in the sulcus instead of the capsular bag. If the same power implant is used, the eye will end up around -1.00 more myopic than expected due to the more forward placement of the IOL.

- If the capsular rupture is accompanied by vitreous loss and a foldable IOL can no longer be used, the surgeon might elect to insert a rigid PMMA implant. If the same power IOL is inserted in the ciliary sulcus, the eye will end up around -2.00 to -2.50 more myopic than expected. This is due not only to the forward placement of the IOL, but also to the fact that most PMMA implants are plano-convex with a more anteriorly located optical center than the biconvex foldable implants.

- If the capsular rupture is so large that the anterior capsular rim can no longer be identified, the surgeon might elect to use an anterior chamber implant. If the same power IOL is used, the eye will end up around -3.00 more myopic than expected. This is due to the forward placement of a PMMA plano-convex implant instead of the normal more posteriorly placed biconvex implant.

Clinical Application

Surgery is planned and IOL calculations call for a 22 D acrylic posterior chamber implant for emmetropia. During surgery, the capsule breaks and vitreous is lost. The surgeon now has to change plans and no calculations for another implant have been made. A decision on the implant power has to be made on the spot. This is where the A-constant comes in handy, no matter what formula has been used.

1. The surgeon decides to insert the same implant in the ciliary sulcus.

When an implant is inserted in the ciliary sulcus instead of the capsular bag, it sits approximately 0.50 to 0.75 mm more anteriorly and each mm change in the ACD affects the final refraction by 1.50 D. To avoid an induced postoperative myopia, an implant that is 1 D less powerful has to be used. In our case, a 21 D acrylic posterior chamber implant should be used.

The difference is less pronounced in long myopic eyes. For example, if a 10 D implant is planned for emmetropia, the surgeon should decrease the power by only 0.50 D and use a 9.50 D implant.

The difference is more pronounced in short hypermetropic eyes. For example, if a 28 D implant is planned for emmetropia, the surgeon should decrease the power by 1.50 D and use a 26.50 D implant.

Clinical Application, continued

2. The surgeon decides to insert a PMMA implant in the sulcus.

Most PMMA implants have plano-convex optics with a more anteriorly located optical center with a different A-constant. In average cases, the power of the IOL lens for emmetropia varies in a one-to-one relationship with the A-constant; if A decreases by 1, the P decreases by 1 D also.

- A-constant of the acrylic posterior chamber lens placed in the bag = 118.4
- A-constant of the PMMA posterior chamber lens placed in the sulcus = 116.4

The difference between the two A-constants is almost 2, which means that the power of the PMMA implant has to be decreased by 2 D. In our case, a 20 D implant has to be used.

The difference is less pronounced in long myopic eyes. For example, if a 10 D implant is planned for emmetropia, decrease the power by only 1.50 D and use an 8.50 D PMMA implant.

The difference is more pronounced in short hypermetropic eyes. For example, if a 28 D implant is planned for emmetropia, decrease the power by 2.50 D and use a 25.50 D implant.

3. The surgeon decides to use an anterior chamber IOL.

Most PMMA anterior chamber implants are plano-convex with a more anteriorly located optical center and a much lower A-constant. In average cases, the power of the IOL lens for emmetropia varies in a one-to-one relationship with the A-constant; if A decreases by 1, the P decreases by 1 D also.

- A-constant of the acrylic posterior chamber lens placed in the bag = 118.4
- A-constant of the PMMA anterior chamber lens placed in front of the iris = 115.3

The difference between the two A-constants is over 3, which means that the power of the PMMA anterior chamber implant has to be decreased by 3 D. In our case, a 19 D PMMA anterior chamber implant has to be used.

The difference is less pronounced in long myopic eyes. For example, if a 10 D implant is planned for emmetropia, decrease the power by only 2.0 D and use an 8.00 D PMMA anterior chamber implant.

The difference is more pronounced in short hypermetropic eyes. For example, if a 28 D implant is planned for emmetropia, decrease the power by 3.50 D and use a 24.50 D implant.

IOL Power Selection in Children

At birth, the child's eye measures approximately 15 mm. The AL increases rapidly to reach 21 mm by the age of 2 years. The growth continues afterward at a much slower rate to average 23.5 mm in adulthood. Similarly, the cornea flattens from an average of 51 D to an average of 44 D between birth and the age of 2 years; the cornea flattens an additional 0.50 D to adulthood.[20]

If surgery is done within the first 2 years of life, and an implant is inserted, a large myopic shift is to be expected a few years later. This has led many surgeons to remove the cataract and fit the eye with a contact lens instead of inserting an IOL.

When surgery is performed after the age of 2 years, there will be a myopic shift as the AL increases with time. This shift will range from 4 to 6 D. Many surgeons recommend undercorrecting the IOL power by around 3 D to partially compensate for the myopic shift; any greater undercorrection would lead to anisometropia and would conflict with amblyopia treatment. The residual myopia in adulthood can be treated with spectacles, contact lenses, or corneal surgery.[21,22]

AL measurement might be a challenge in these young patients. If this is the case, it is recommended to measure the corneal power and the AL under general anesthesia.

References

1. Binkhorst RD. Biometric a-scan ultrasonography and intraocular lens power calculations. In: Emery JM, ed. *Current Concepts in Cataract Surgery: Selected Proceedings of the Fifth Biennial Cataract Surgical Congress.* St. Louis, MO: The CV Mosby Co.; 1978: 175-182.
2. Boerner CF, Thrasher BH. Results of monovision correction in bilateral pseudophakes. *American Intra-Ocular Implant Society Journal.* 1984;10:49-50.
3. Drews RC. A practical approach to lens implant power. *American Intra-Ocular Implant Society Journal.* 1977;3:170-176.
4. Drews RC. The determination of lens implant power. *Ophthalmic Surg.* 1989;20:625-637.
5. Fechner PU, Kania J, Kienzle S. The value of a zero power intraocular lens. *J Cataract Refract Surg.* 1988;14:436-440.
6. Gill JP. Minimizing postoperative refractive error. *Contact Intraocular Lens Med J.* 1980;6:56-59.
7. Hoffer KJ. Preoperative cataract evaluation: intraocular lens power calculation. *Int Ophthalmol Clin.* 1982;22:37-75.
8. Menezo JL, Cisneros A, Harto M. Extracapsular cataract extraction and implantation of a low power lens for high myopia. *J Cataract Refract Surg.* 1988;14:409-412.
9. Shammas HJ. Spectacle correction desired after cataract removal. *J Cataract Refract Surg.* 1991;17:101-102.
10. Kora Y, Yagushi S, Inatomi M, Ozawa, T. Preferred postoperative refraction after cataract surgery for high myopia. *J Cataract Refract Surg.* 1955;21:35-38.
11. Huber C, Binkhorst C. Iseikonic lens implantation in anisometropia. *American Intra-Ocular Implant Society Journal.* 1979;5:194-198.
12. Miyaki S, Awaya S, Miyaki K. Aniseikonia in patients with a unilateral artificial lens, measured with Aulhorn's phase difference haploscope. *American Intra-Ocular Implant Society Journal.* 1981;7:36.
13. Troutman RC. Artiphakia and aniseikonia. *Amer J Ophthalmol.* 1963;56:602-639.
14. van der Heijde GL. The optical correction of unilateral aphakia. *Trans Amer Acad Ophthalmol Otolaryngol.* 1976;81:80-88.
15. Binkhorst RD. Pitfalls in the determination of intraocular lens power without ultrasound. *Ophthalmic Surg.* 1976;76:69-82.

16. Holladay JT, Rubin ML. Avoiding refractive problems in cataract surgery. *Surv Ophthalmol*. 1988;32:357-360.
17. Hoffer KJ. Biometry of 7500 cataractous eyes. *Amer J Ophthalmol*. 1980;90:360-368.
18. Kratz RP, Shammas HJ. *Color Atlas of Ophthalmic Surgery. Cataracts*. Philadelphia, PA: JB Lippincott Co.; 1991: 13-19.
19. Shammas HJ. *Atlas of Ophthalmic Ultrasonography and Biometry*. St. Louis, MO: CV Mosby Co.; 1984: 273-308.
20. Gordon RA, Donzis PB. Refractive development of the human eye. *Arch Ophthalmol*. 1985;103:785-789.
21. Sinskey RM, Patel J. Posterior chamber intraocular lens implants in children: report of a series. *American Intra-Ocular Implant Society Journal*. 1983;9:157-160.
22. Vasavada A, Chauhan H. Intraocular lens implantation in infants with congenital cataracts. *J Cataract Refract Surg*. 1994;20:592-598.

The Unexpected Postoperative Result

With the refinement of intraocular lens (IOL) formulas, patients and surgeons are now expecting a more accurate final refraction in the operated eye. Twenty-five years ago, results within ± 3.0 diopters (D) were acceptable and deemed superior to any aphakic correction.[1-4] Nowadays, our results are accurate to within ± 0.5 D.

All available IOL formulas are based on mathematical equations with modifications obtained from clinical results. They all require three entries: the axial length (AL), the corneal power, and a constant ("A" constant in the SRK/T formula, "S" factor in the Holladay formula, or the "ACD" value in the Hoffer Q formula). An error in the determination of any of these entries will undoubtedly result in an unexpected postoperative result.

Occasionally, the error is very large and exceeds 5 D. However, in most cases, the error ranges between 2 and 5 D, resulting in an induced myopia or hyperopia in the operated eye.

The Large Error

When the error is very large and exceeds 5 D,[5-8] the IOL has to be exchanged. The most common causes include the following.

A Large Error in Axial Length Measurement in the Very Long or Very Short Eye

The most common condition that induces an error in AL measurement is an unsuspected unilateral axial myopia.[8] This is a rare congenital anomaly that causes anisometropia and amblyopia in the affected eye. It remains undetected throughout childhood. Patients, usually in their 40s and 50s, present to the ophthalmologist

when they lose vision in the affected eye because of the cataract's progression. When the patient is examined for the first time, the myopia is not detected for two reasons:
1. The patient is not wearing the myopic correction because of the associated anisometropia and amblyopia.
2. The preoperative refraction is unreliable because of the opaque media.

Unilateral axial myopia is the most common cause of large errors.

HUMAN ERROR

Human error is usually at the base of the inadvertent use of a different power IOL. A patient with a long history of emmetropia is expected to need an 18 to 22 D implant, while a long-time myope is expected to need less than 18 D and a hyperope more than 22 D. If the calculations call for a drastically different power implant, the calculations should be repeated and confirmed prior to surgery.

THE USE OF MISLABELED INTRAOCULAR LENSES

In the early days of lens implantation, surgeons encountered some cases of mislabeled IOLs due to the fact that some IOL companies did not use adequate quality control during the manufacturing process.

Induced Postoperative Myopia

When confronted with an unexpected postoperative myopia (-1.00 to -5.00 D), the following steps are followed in search of the cause:[9-21]
1. The power of the inserted IOL is rechecked against the calculated power to rule out human error. To avoid such a mistake, the surgeon should always personally check the power of the IOL prior to inserting it in the eye.
2. Both eyes are remeasured to rule out an erroneous preoperative shorter AL measurement, which is the most common cause of error in the presence of an induced myopia. The causes of such a shorter measurement and how to avoid them are detailed in Chapter 13. The use of a shorter AL measurement in the IOL power calculations will call for the use of a stronger IOL than is actually required, resulting in an induced myopia in the final postoperative refraction.

An erroneous shorter AL measurement is the most common cause of induced postoperative myopia.

3. The corneal power is recalculated to rule out a postoperative steepening of the cornea. It is extremely unusual for the corneal power to change after a routine phacoemulsification cataract extraction. Occasionally, tight sutures placed at 12 o'clock will induce with-the-rule astigmatism and corneal steepening in the vertical axis; suture removal will restore the preoperative corneal curvature. Induced myopia due to corneal steepening is more likely to occur after a triple procedure of corneal transplant-ECCE-lens implantation, or of corneal trans-

plant with IOL removal and replacement. Induced myopia after a triple procedure can be avoided by using a corneal button that is only 0.2 mm larger than the recipient, instead of a 0.5 mm, and performing a 7.5 to 8.0 mm transplant instead of a 7.0 mm transplant.

4. The implant's position is reassessed to rule out a forward placement with or without tilting of the IOL within the eye. During ECCE surgery, the posterior chamber implant is usually placed in the capsular bag. However, in certain cases, the implant is inserted in the ciliary sulcus or will have one of its loops in the bag and the second loop in the ciliary sulcus. The implant will tilt inside the eye and sit closer to the cornea than expected, causing an induced myopia. The condition is exaggerated if the anterior chamber shallows postoperatively due to a wound leak or a concomitant filtering surgery.

Induced Postoperative Hyperopia

When confronted with an unexpected postoperative hyperopia, the following steps are followed in search of the cause:[9-21]

1. The power of the inserted IOL is rechecked against the calculated power to rule out human error. To avoid such a mistake, the surgeon should always personally check the power of the IOL prior to inserting it in the eye.

2. Both eyes are remeasured to rule out an erroneous preoperative longer AL measurement. A longer AL can occasionally be measured prior to cataract surgery. The causes of such a longer measurement are detailed in Chapter 13. The use of a longer AL measurement in IOL power calculations will call for the use of a weaker IOL than is actually required, resulting in an induced hyperopia in the final postoperative refraction.

3. The corneal power is recalculated to rule out a postoperative flattening of the cornea. Postoperative corneal flattening is a rare occurrence after a routine extracapsular cataract extraction. Occasionally a wound gape flattens the cornea in one meridian. Repair of the wound gape corrects the condition. Also, corneal flattening has also been noted as a complication of cataract surgery after corneal refractive surgery. This condition is discussed in detail in Chapter 17.

4. In short eyes, the IOL formulas are rechecked for accuracy. The use of the SRK and SRK II equations in the short eye will call for a much weaker implant, resulting in an induced hypermetropia.[11] The original SRK and the modified SRK II equations have been replaced by the more accurate SRK/T formula. However, the SRK and the SRK II equations are still being used, to date, by some surgeons, especially in some Third World countries.

The use of the SRK and SRK II formulas is the most common cause of induced postoperative hyperopia in short eyes.

Correcting the Postoperative Refractive Error

Patients' reactions vary according to their expectations. Patients have a tendency to complain more if the error occurs in the dominant eye with an unexpected anisometropia and/or aniseikonia. In myopes and hyperopes that have always worn glasses for distance vision, an error in IOL power calculations is covered up by a postoperative change in the lens prescription and does not always result in a postoperative disappointment to the patient. Emmetropes, on the other hand, have higher expectations; an induced error in the dominant eye with anisometropia and/or aniseikonia will certainly result in patient disappointment. In such cases, the surgeon and the patient have multiple choices:

1. Fit the operated eye with glasses or a contact lens.
2. Fit the nonoperated eye with a contact lens.
3. Corneal refractive surgery.
4. Exchange the implant.
5. Insert a piggy-back implant.

These choices are discussed with the patient and the appropriate course of action is taken.

References

1. Binkhorst RD. Pitfalls in the determination of intraocular lens power without ultrasound. *Ophthalmic Surg*. 1976;76:69-82.
2. Clevenger CE. Clinical prediction versus ultrasound measurement of IOL power. *American Intra-Ocular Implant Society Journal*. 1978;4:222-224.
3. Kraff MD, Sanders DR, Lieberman HL. Determination of intraocular lens power: a comparison with and without ultrasound. *Ophthalmic Surg*. 1978;9:81-84.
4. Olson RJ. Intraocular lens power calculations: an extra edge or expensive waste? *Arch Ophthalmol*. 1987;105:1035-1036.
5. Salz JJ, Reader AL III. Lens implant exchanges for incorrect power: results of an informal survey. *J Cataract Refractive Surg*. 1988;14:221-224.
6. Shammas HJ. The "9 diopter" surprise revisited. *J Cataract Refract Surg*. 1988;14:580.
7. Shammas HJ. Axial length measurements and IOL power calculations in microphthalmic eyes. In: Sampaolesi R, ed. *Ophthalmic Ultrasonography: Proceedings of the 12th SIDUO Congress*. Documenta Ophthalmologica Series. Dordrecht, The Netherlands: Kluwer Academic Publishers; 1990; 53:145-148.
8. Shammas HJ, Milkie, CF. Mature cataracts in eyes with unilateral axial myopia. *J Cataract Refract Surg*. 1989;15:308-311.
9. Cravy T. Using the intraocular lens refraction factor to improve refractive prediction accuracy. *J Cataract Refractive Surg*. 1989;15:519-525.
10. Hillman JS. Intraocular lens power calculation for emmetropia: a clinical study. *Br J Ophthalmol*. 1982;66:53-56.
11. Hoffer KJ. Intraocular lens calculations: the problem of the short eye. *Ophthalmic Surg*. 1981;12:269-272.
12. Holladay JT, Prager TC, Long SA, Koester CJ. Determining intraocular lens power within the eye. *American Intra-Ocular Implant Society Journal*. 1985;11:353-363.
13. Holladay JT, Rubin ML. Avoiding refractive problems in cataract surgery. *Survey Ophthalmology*. 1988;32:357-360.
14. Hoffer KJ. Accuracy of ultrasound intraocular lens calculation. *Arch Ophthalmol*. 1981;99:1819-1823.

15. Huber C. Effectiveness of intraocular lens calculation in high ametropia. *J Cataract Refract Surg*. 1989;15:667-672.
16. Olsen T. Sources of error in intraocular lens power calculations. *J Cataract Refract Surg*. 1992;18:125-129.
17. Olson RJ, Kolodner H, Kaufman HE. The optical quality of currently manufactured intraocular lenses. *Am J Ophthalmol*. 1979;88:548-555.
18. Phillips P, Perez-Emmanuelli J, Rosskothen HD, Koester CJ. Measurement of intraocular lens decentration and tilt in vivo. *J Cataract Refract Surg*. 1988;14:129-135.
19. Richards SC, Olson RJ, Richards WL. Factors associated with poor predictability by intraocular lens calculation formulas. *Arch Ophthalmol*. 1985;103:515-518.
20. Shammas HJ. Accuracy of lens power calculations with biconvex and meniscus intraocular lenses. *Am J Ophthalmol*. 1988;106:613-615.
21. Shammas HJ. *Atlas of Ophthalmic Ultrasonography and Biometry*. St. Louis, MO: CV Mosby Co.; 1984: 273-308.

Index